THE PHYSICS OF FILTER COFFEE

Jonathan Gagné

to Amélie and Mélinda

Text by Jonathan Gagné

Editing by Jean Zimmer

Photography by Noé Aubin-Cadot, Kathy Gagné, Dylan Siemens, Elika Liftee, Andreas Corcaci and Jonathan Gagné

Logo on front cover by Daniel Demay

Book and cover design by Rebecca S. Neimark, Twenty-Six Letters
http://www.twentysixletters.com

Published by Scott Rao

Printed in China

ISBN 978-0-578-24608-6

Jonathan's blog, website, and contact information: http://www.coffeeadastra.com

To purchase this book or make a wholesale inquiry, please visit http://www.scottrao.com

Table of Contents

Foreword

There is a long list of people without whom I would never have gathered the knowledge or motivation to write this book. But first, none of this would have been possible without my wife Amélie's boundless patience while I cluttered our kitchen and apartment with vacuum-sealed coffee, heaps of coffee gear, and weird tools.

I am not sure that I would have even discovered my love for coffee if I had not first encountered coffee prepared in a way that opened my mind to the flavor complexity it can have, and for that I have to thank many masterful baristas around the world, such as Chris Capell and Sandra Wolter. But the person who most clearly sparked this journey into the physics of coffee is certainly Scott Rao, first through his books and then through his early encouragements when I began writing about coffee online. Our endless coffee discussions, where my naive intuitions of physics met with his decades of experience, made me progress at a rate that would have otherwise been impossible.

I am also especially grateful to Matt Perger and the Barista Hustle team, who encouraged me and with whom I have also had many exciting discussions, to Andy Kyres for generously providing me with lots of coffee, ample space, and tools to experiment and season grinders at the Canadian Roasting Society, and to Dan Eils for 3D printing and sending me almost every bad idea I had. Francisco Quijano provided help I could not overstate by writing web applications to gather data from Bluetooth scales, which would otherwise have taken me months of harrowing work. Thanks to Edo Rezaprasga for helping me put my hands on obscure coffee gear only available in Asia and to Alex Caza and everyone at the Firstbloom app for generously sharing their amazing data with me. Thank you, Noé Aubin-Cadot and Kathy Gagné, for providing photos and Daniel Demay, for creating a logo for this project. Thank you Jean Zimmer and Rebecca Neimark for your invaluable input that made this book so much better.

I would also like to thank Christoper Feran, Chris Capell of the Montreal Coffee Academy, Samo Smrke, Kevin Moroney, John Buckman of Decent Espresso, Wille Yli-Luoma of Heart Coffee Roasters, Doug Weber of Weber Workshops, Mitch Hale, Ray Murakawa of Melodripco, Andreas Corcaci, Dany Marquis of Brûlerie du Quai, Umeko Motoyoshi, Jonathan Franklin, Matt Leberman, Stéphane Ribes, Catherine Gauthier, and Trent Howard for sharing their knowledge with me and for interesting discussions or ideas. James Hoffmann and Tim Wendelboe put discussions out in the coffee community that helped shake out my thoughts and generate new ideas. Thanks to Caleb Fischer for sending me an insane amount of rare coffees and to the many others who also shared with me more coffee than what could fit in my freezer: Alex Levitt, Kalyana Vattikuti, and Andreas Corcaci.

I would also like to thank many who directly encouraged me in this coffee quest: Vince Fedele, Sébastien Blondeau, and Elise Arsenault from the Café Normal podcast, Thibaut Paggen from Café Pista, Magalie "Steve" Rondeau, Patrick Liu, David Boucher from Escape Coffee Roaster, David Lalonde from Rabbit Hole Roasters, Max Roderbourg, Dave Wong, Sherry Morrison, Jeremy from Socratic Coffee, Anne Roger, Max Charron, Dylan Siemens, and Elika Liftee.

Pursuing this goal of a better understanding of coffee would have been so much slower without the help that Patreon followers brought me to purchase various high-quality measuring or data-collection devices. I am extremely grateful to every one of you: Michael Replan, Francisco Quijano Villanueva, Adnan Fakhry, Joseph Jang, Matthew Brown, Gwilym Davies, Lazaros Sotiropoulos, Andrew Baker, Lawrence Yeh, Erick Leiva, Jesse Lambeth, Sioban Staplin, Tony Hariman, Roger Yoon, Kyle Armstrong, Eric Himanga, William Barry, David Ramsey, Camilo Velez, Friedrich Scheck, Dany Marquis, Benjamin Glennon, Nik Reese, Chakorn Chamnanchang, Matt Chitharanjan, Simon He, Elmar Pfeilsticker, Clay Bush, Roger Jordan, Lee Devin, Matthew A., Emilio Rodriguez Rebollo, Wipop Suppipat, Benjamin Chia, Chris Lehmann, Elmar Franke, Patrick Stanger, Matthew Hoadley, Brendan Ojala, Brett Turner, Thomas Dunn, Aurelien He, Merrick, Chase Foster, Suwitcha Chandhorn, Alan Ho, Garam Um, Pieter Thyssen, Matteo Cecconi, Milorad Sekulovic, Patrick Snook, Steve Bercovici, Kyle Tubbs, Camilo Cortes, Simon Mathers, Diego, Steven McLoughlin, Leoniel Anthony De Leon, Thibaut Paggen, Tarik Tahir, Okan Turfanda, Eric Nadler, Ghazi Almoayed, Ignasi Notario, James Hoffmann, Viktor Bobek, Ray Murakawa, Dan Eils, Joachim Leong Ming Yoong, TopDownTom, Mustafa Aydin, Brook Stefanou, Răzvan Roșu, Jasmine Segura, Anders Christiansen, Sebastian Andersson, Siarhei Laurenau, Ron Donaldson, Norman Mazel, Niklas Kügler, Max Roderbourg,

Manu Taro, Tam Yin Ping, David Yardley, Craig Batory, Andrea Beckham, Regan Wood, Tudor Cirstea, Trym Aakre, Thiago Cavalheiro, Boris Lee, Dan Davis, Mumin Bayoumi, Specialty Coffee College, Ali Abdullah Aldiwani, Roman Vilanov, Yosho Tsuchiya, Thomas Greene, Chad Brizendine, Patrik Breitenmoser, Khian Le, Aart Eskes, Anthony Nguyen, Luca Giorgella, Jaroslaw Wawro, Marius Disler, Joe Lin, Takumi Nakata, James Howell, Heath Dalziel, Sal C., Fabian Schläpfer, Daniel Ehrlich, Tonx Konecny, Adam Holliefield, Dorian Arcos, Артём Шрам, Oleg Patrakov, Felix Julian Frank, Dalton Robison, Nathaly, Abdulmajeed Alahmadi, Adrian Hasley, Don Cho, Ryan McDonnell, Rohit Sud, Rana Hassanieh, Anthonia Mao, Brady McAtee, Kai Gibson, Shawn Adkins, Jayce Van Der Linden, Sarah J. Hannaway, Daniel Gutierrez, Achim Klein, Connor Garland, Kevin Davis, Jay Kim, Sorin Lupu, Terça Expressa, Zack Stamey, Gonçalo Ávila, Felix, Rohan Bhattacharya, Kahvecinin Galaksi Rehberi, Morteza Bagherpanah, Rafael Cascao, Qasim Nathoo, Aurore Vayer, Mihai Berbec, Edison Ting, Nuno C. Sousa, Octavio H. Ruiz Cervera, Smag, Matthias Harlass, Jeehae Ahn, Brook Stefanou, Joseph O'Keefe, Emil, Daniel Cobas, Tamer Hourani, Can Gencer, Cameron Sawyer, Robert Hall, Travis Muzzillo, Andrea Villani, Ali Yousefi, Alexey Mordovin, Manuel Jesús Gutiérrez Torres, Bushi Drop, Suksit Theparee, Laszlo Geszten, Nasko Panov, Andrew Zaporozhets, Adrian Bularca, Alex Dres, Adam Fortais, Moritz Rzehak, Paul Arnephy, Brennan Salte, Philippe Gauthier, Jesse Beahler, Le Doan Thanh Trung, Kasjan Orzol, Frédéric Gabioud, Federico Bolanos, Philip Szobody, John Jacobson, Tim Piazza, Cole Hladun, Brandon Carbullido, Sergei Kutrovski, Michael Plunkett, 洪强 张, Jiri Brazdil, Serkan Kenar, Benji Schuetz, Michael F. Stewart, Henry Watson, Alice Yan, Joaquin Lencinas, Joseph O'Tweefe, Jonas Schlaepfer, Jeppe Hagedorn, Ian Chan, Walter Enomoto, and Gregory Bachman.

I would also like to thank some companies that offered either free material or rebates on their products to encourage my research. Those include VST labs, Fellow Products, Weber Workshops, Brewcoat, and Lilydrip.

Avoiding Conflicts of Interest

At the time of writing this book, the author Jonathan Gagné has no financial ties or interests to any coffee-related companies or institutions. The companies who provided direct benefits in the form of material or discounts on purchases are listed in the paragraph above.

Introduction

My goal in writing this book is to provide readers with a mental toolkit to better understand filter-coffee brewing and how we can affect it. I do not wish to convince readers about what method is best, what kettle temperature we should use, or what type of grinder is the correct one for filter coffee. Because coffee tasting is subjective, it is likely that different people will prefer different methods, and I am honestly not sure that I could even predict what brew method I would prefer the most among the innumerable possibilities.

Coffee brewing is a topic of surprising complexity that defied my predictions many times and seemed like a matter of black magic for years. I have often witnessed contradicting explanations for an observed effect, some more scientific than others. The more I learned about coffee brewing, the more I realized how we seem to have discovered the most popular tools and methods almost by random exploration. I do not feel judgmental about this, however; random exploration is about the best we can do when we do not have the knowledge to crack open a black box. I hope this book will contain some of this knowledge and help you explore the possibilities of coffee brewing in a more efficient way, whatever your goal is.

Early on, when I started on this journey down the rabbit hole of coffee physics, I put together a blog at *coffeeadastra.com*, where I wrote about concepts of coffee physics as I was discovering them, hoping I could synthesize and share this knowledge with the geekiest of coffee geeks. This audience turned out to be much larger than I had expected. The extremely positive response and the tremendous help I encountered from the coffee community encouraged me to delve even deeper, almost to the point of turning my kitchen into a full-fledged laboratory. Only recently have I begun to understand the physics of coffee extraction enough to feel confident about writing a book on the topic.

This book will focus on the physics of filter coffee brewing, with particular attention to manual *percolation* methods such as pourovers. This is the first angle from which I began exploring the concepts underlying coffee extraction, simply because this turned out to be my favorite brew method. Some of you may be more interested in espresso, or even in the chemical composition of coffee and how it interfaces with sensory experience; you might find interesting bits of information spread through this book, but please keep in mind that these topics are not my main focus, nor are they anywhere close to my area of expertise.

The brewing of a pourover coffee, where hot water is manually poured into a coffee *dripper*

I wrote this book with the intention of providing tools for coffee professionals and home baristas who have no prior knowledge of physics or mathematics. We will at times delve into relatively simple equations, but I will take the time to explain every term, with the aim of making the equations illustrative rather than intimidating. Please bear in mind that if some concepts go above your head or seem way too complicated, it will have been my failure to make them clear enough, not your failure of understanding. I hope that, even if this happens, it will not make the remainder of the book more opaque.

My primary work experience is in the field of physics, specifically astrophysics. Along the journey to becoming a professional researcher in this field, I learned mostly about physics and very little about chemistry and biology—these two fields of science bear tremendous impact on our understanding of coffee, but I can hardly contribute to these aspects. I do not even consider myself an expert on the topic of fluid dynamics, which sits squarely within physics, and I am certainly not an expert on percolation—my main expertise is much closer to poorly structured computer programming, data visualization, statistics, and the very empirical gathering of astronomical data at the telescope. Somehow, these skills still turned out to be useful in my quest to better understand coffee brewing.

This book is structured in chapters, each of which covers an important aspect of brewing filter coffee, from extraction, water, grinders, and percolation to filters, kettles, and dripper design. I show how coffee beans and roasting practices affect brewing, and I present practical recommendations that can be applied to brewing based on the physics described in the first chapters. I end with a description of different instruments relevant to gathering coffee data and how I used them.

Glossary

Readers who are not familiar with coffee jargon could get confused or distracted by some terms I use throughout this book. I therefore recommend reading this glossary first if you have not already read technical books or blogs about coffee extraction.

AeroPress: A piston-like plastic brewer manufactured by Aerobie in the USA.

Astringency: A sensation of constriction on the tongue often associated with an over extraction of coffee solubles.

Average extraction yield: The fraction of the dry coffee bed, by weight, that was dissolved in water.

Beverage weight: The total weight of brewed coffee to be consumed.

Bloom: A step of preinfusion that allows the coffee to be wetted and to begin degassing and absorbing water before the main percolation phase.

Body: The body of a coffee beverage is related to its mouthfeel. A coffee with more body feels heavier, and seems to have more texture.

Brew ratio: The ratio of the total weight of water used for coffee preparation over the weight of the dry coffee dose.

Bypass: The act of water passing around the coffee bed, thus reducing the extraction of coffee solubles.

Channels: Regions of lower hydraulic resistance in the coffee bed where water flows faster, resulting in local overextraction.

Coffee dose: The weight of dry coffee from which the beverage was prepared.

Colloids: Undissolved solids in suspension in a coffee brew.

Concentration or TDS: The fraction of a coffee beverage that is made of dissolved coffee solubles, by mass. TDS stands for total dissolved solids.

Dial-in: Adjusting one's coffee tools or recipe to achieve the desired coffee. This often refers to grind size specifically.

Drawdown: The moment at which water has completely drained out of a coffee dripper.

Dripper: A device that holds a filter and dry coffee grounds to achieve percolation.

Gooseneck: The thin, elongated spout of a precision kettle that allows the user to better control the flow rate.

Hard water: Water with a high mineral content.

Hydraulic conductivity: A measure of the tendency of a specific coffee bed, water, and local gravity in producing a high flow rate. This includes the properties of water, such as its temperature-dependent viscosity. Hydraulic conductivity is the inverse of hydraulic resistance.

Hydraulic permeability: The intrinsic capacity of a coffee bed to let fluids flow through, in general.

Hydraulic resistance: A measure of the tendency of a specific coffee bed, water, and local gravity to slow down the flow of water; i.e., the inverse of hydraulic conductivity.

Immersion brew: A method of coffee preparation wherein ground coffee is added into still water for a set amount of time.

Liquid retained ratio: The ratio of the weight of water that was retained by the spent coffee bed over the weight of dry coffee.

Particle size distribution: the fraction of ground coffee particles of each size as produced by a grinder.

Percolation: The action of letting water drip through a bed of coffee to achieve both extraction of coffee solubles and filtration of undissolved solids.

Permanent hardness: The remaining mineral content of water if it is allowed to form as much precipitate (usually as deposits of calcium carbonate) so as to deplete it completely of its bicarbonate ions.

Precipitate: A solid deposit formed by dissolved ions that recombined and came out of solution.

Quakers: Unusually pale coffee beans that can impart a negative taste on a brew.

Refractometer: An instrument that measures the refractive index of a liquid, i.e., the angle by which light is deflected when it passes through the medium.

Server: The container into which the extracted coffee beverage falls during percolation.

Slurry: The mix of ground coffee and hot water in an actively extracting coffee preparation.

Soft water: Water with a low mineral content.

Soluble: A chemical compound that can be dissolved in water.

Swirl: The act of moving a dripper along in small circles horizontally.

TDS: Total dissolved solids, or the concentration of solubles dissolved in water, in fractional mass units.

Temporary hardness: The amount of mineral content in water that would be able to form a precipitate, usually by combining with the same amount of bicarbonate (when defining amount as a number of ions).

Titration: A method by which a chemical is added to a solution to react with a specific ion and measure its concentration by observing a color shift when the target ion is completely reacted away.

Total alkalinity: The capacity of water to resist changes in pH as acids are added to it. In the context of coffee brewing, this is the concentration of bicarbonate (HCO_3^-) ions.

Total hardness: The total mineral content of water, usually made of positive magnesium and calcium ions.

V60: A conical dripper manufactured by Hario in Japan, that features walls angled at 60° from the horizontal.

Variety/varietal: An ensemble of plants possessing similar genetic and phenotypic properties. A variety is more specific than a species or subspecies. The term "varietal" is often incorrectly used to specify a variety in the context of coffee.

Water weight: The total weight of water that was poured over a dry coffee bed in a percolation. This includes both the water that ended up in the beverage and the water that was retained in the spent coffee bed.

CHAPTER 1

Extraction

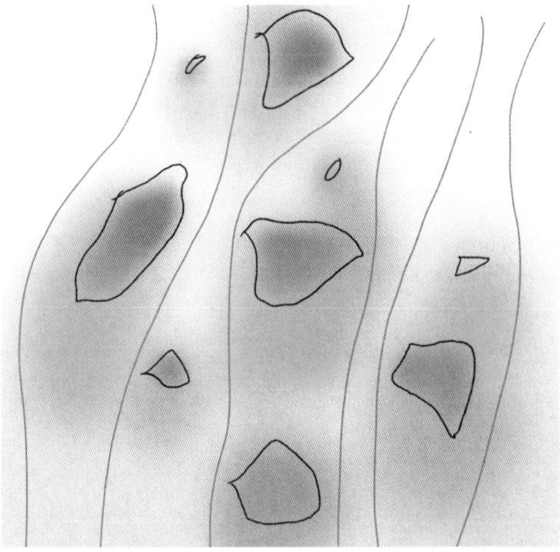

Before we can have a meaningful discussion about how every aspect of coffee brewing affects the resulting cup of coffee we drink, it will be useful to acquire a basic understanding of how water extracts the chemicals responsible for the taste of coffee. In this chapter, we review some of the chemical and physical phenomena responsible for extraction. We also talk about some practical concepts such as the notions of *concentration* and *average extraction yield,* how to measure and calculate them, and why they are useful.

We will not go deep into the detailed chemistry of coffee compounds, since they are extremely complex and rich. For more discussion on the topic of coffee chemistry, I recommend the books *Coffee: Production, Quality and Chemistry* (Farah 2019); *The Craft and Science of Coffee* (Folmer 2017); and *Espresso Coffee: The Science of Quality* (Illy and Viani 2004).

1.1 Average Extraction Yield

One of the most useful concepts in coffee brewing is the average extraction yield. Put simply, it represents the mass fraction of the dry coffee that was extracted and dissolved into your cup. A large fraction of the coffee beans cannot be dissolved in water. For example, the cellulose that makes up the walls of the coffee cells at the microscopic level is not soluble in water. Because of this, only about

30–32% of typical coffee beans (by weight) can really be extracted with standard brew methods (e.g., de Maria et al. 1994; Ellero and Navarini 2019), and that puts an upper limit on the highest average extraction yields we might encounter.

We call this the "average extraction yield" and not just "extraction yield" because not all coffee cells usually extract to the same level. Consequently, your cup of coffee likely contains the *solubles* from a collection of some poorly extracted cells and some highly extracted cells, as well as others in between. The average of all these extraction levels will determine your average extraction yield. The concept of average extraction yield is different from that of concentration. The concentration of a coffee drink depends not only on the quantity of compounds you have extracted from the dry coffee (i.e., the average extraction yield), but also on how much water you have diluted them in.

Putting the concepts of average extraction yield and coffee concentration together, Earl E. Lockhart created the Coffee Brewing Control Chart in the 1950s at the Coffee Brewing Institute.

Through multiple experiments and collection of sensory data from the general public, the Coffee Brewing Institute and, later, the Norwegian Coffee

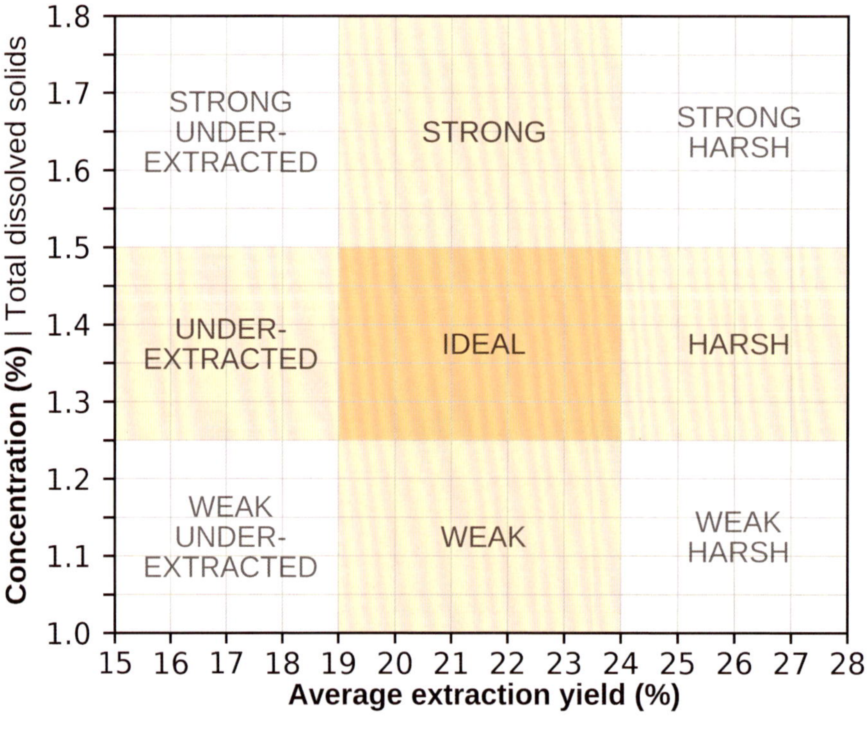

An updated version of the original Coffee Brewing Control Chart by Lockhart. The average extraction yield (horizontal axis) correlates with flavor profile, and concentration (vertical axis) correlates with flavor intensity.

Association determined consistent optimal zones in the Coffee Brewing Control Chart that seemed to be preferred by most coffee drinkers (Lingle 1996). Average extraction yields below 18% were deemed underextracted and associated with less-pleasant flavors such as grass and peanuts, and those overextracted above 22% were associated with bitterness and *astringency.* Since these early developments, it has become clearer that the optimum average extraction yield is significantly impacted by factors such as roasting (Rao 2014), grinder performance (Rao 2018), brewing methods, and green coffee quality, as well as personal preference. We will explore these effects in more details in the following chapters of this book, but it is useful to keep in mind that recent developments in roasting, farming, and grinder quality have led the typical range of preferred average extraction yields to shift slightly toward higher values.

While this is not based on a statistical population analysis, average extraction yields below 19% are nowadays less often described as "preferred," and some influencers in the coffee industry have noted a preference for average extraction yields well above 22%. Only in rare cases have I personally preferred coffees brewed with average extraction yields above 23.5%; even when using a high-quality grinder, the quality of green coffee and roasting appear to play a very important role in the optimum window of average extraction yield. It is interesting to note that preferences in terms of average extraction yield do not tend to be linear: for example, Rao (2016) noted that there seems to be an optimum in taste preference near average extraction yields of around 15%, followed by a region that most people dislike (15% to 18%), followed by another positive evolution in preferences past 18%.

The optimum zone in coffee concentration is correlated with flavor intensity rather than the flavor profile, and as such there is a much wider range of preferences. The Coffee Brewing Institute found an acceptable range between 1.0% and 1.5%, but the lower end of this range is now more rarely encountered in the specialty coffee community. These preferences are, in principle, independent of those in average extraction yield. They are simply based on how strong we like our coffee to taste. It is notable that drip coffee is rarely prepared with concentrations in the range 1.8–7.0%, yet espresso coffee is mostly consumed at concentrations of 7.0–12.0% (Perger 2017, Jan. 30) with similar average extraction yields. This illustrates how the preferred window of concentrations can be much wider than that of average extraction yields.

The average extraction yield is a useful concept mostly because it has a large impact on the taste profile of coffee. To understand why this is so, we need to realize that coffee cells contain a large amount of distinct chemical compounds that together build the perceptive experience of coffee flavor. However, they do not all extract at the same rate. Some compounds extract very quickly and others more slowly: because of this, a cup of coffee with an average extraction yield above 28% would include almost every dissolvable compound that the coffee bean contains. Typically, this is not a good thing because some of the slower-extracting compounds can taste harsh, bitter, and/or metallic and cause an astringent sensation on the tongue. Conversely, a coffee extracted at around 15% would exclude a lot of the interesting compounds and would likely lack sweetness and complexity.

This observation provides an underlying explanation for Lockhart's findings and allows us to understand that the average extraction yield is a number that correlates with the *taste profile*. For many reasons, it is by no means a perfect indication of what a coffee will taste like. For example, two cups of coffee, both with an average extraction yield of 20%, can be constituted either of a perfectly even extraction (i.e., all coffee cells are extracted at 20%) or a very uneven cup (where some cells are extracted at 28% and others at 15%). The two resulting cups will undoubtedly taste very distinct from each other. On top of this consideration, two cups of coffee prepared at the same average extraction yield but with different coffee beans, grinders, roast profiles, or even water compositions can also taste different. For these reasons, it is important to keep in mind that the average extraction yield is a tool of limited use, even though it may be the most powerful one at the disposition of baristas to investigate how their actions affect extraction.

One of my main goals in writing this book is to explain how we can affect extraction—specifically, the evenness of extraction—across coffee cells. I do not know whether more-even extractions will always taste better, but in my limited experience I seem to prefer them. Even if we desire some degree of unevenness in our extractions, I think it would be beneficial to produce it in a deliberate way. Not only will this allow us to replicate our results more easily, but it might guide us into unexpected territories in terms of what we can do to improve the taste of coffee.

In summary, the Coffee Brewing Control Chart is a simple yet powerful tool to guide us in how we brew coffee. If a cup of coffee were a musical instrument, the horizontal axis, or average extraction yield, would represent the tone of the instrument, and the vertical axis, or concentration, would represent how loud the instrument is played. At the end of this chapter we review how average extraction yield can be calculated, and in Chapter 11 we detail a method to measure coffee concentration accurately using a *refractometer*.

An example of diffusion: ink slowly spreads out over time. Source: Wikimedia Commons

1.2 Advection and Diffusion

Now that we have discussed the sensory results of extraction, we will turn our attention to the physical mechanisms that allow extraction to happen. We will focus on the case of percolation, wherein water is poured over the coffee grounds and drips through. Two different phenomena in physics can transport dissolved coffee compounds from the inside of the ground coffee into your beverage: advection and diffusion.

Advection is the technical term for simply transporting a compound with the flow of water. You can imagine the coffee compounds like leaves floating on a river; as the water moves along, the compounds are displaced with the current. Advection is often referred to as the "washing out" of coffee solubles.

Diffusion describes how a chemical compound will distribute itself, even in the absence of flow. To understand why this happens, it is useful to dig a little bit into the concept of temperature. The particles that constitute a fluid are constantly moving around along erratic trajectories. The average velocity at which they do so is related to what we call the *temperature* of a fluid. In other words, any fluid contains particles that are constantly shuffling around; the warmer it is, the faster the particles move. Any dissolved chemical compounds will slowly spread out because of these random motions. If we wait long enough, diffusion always moves chemicals from the most concentrated areas to the least concentrated areas, until a balance is reached at which the concentration is the same everywhere. The propagation of smoke in a room free of airflow is a nice illustration of

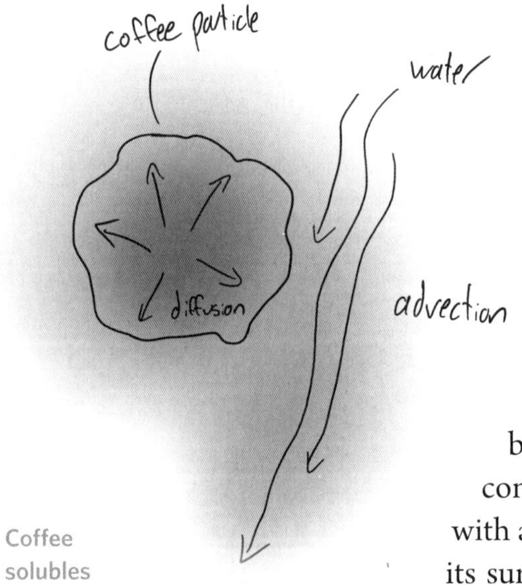

Coffee solubles diffusing to the surface of a coffee particle and subsequently getting advected by the flow of water

diffusion; the initial patterns that may have been created by the source of smoke become blurrier as the smoke particles diffuse outward.

Ellero and Navarini (2019) built a detailed model of how extraction happens at the microscopic level, and they confirmed its validity by comparing it to experimental espresso extraction data. This is the basic idea: a particle of ground coffee consists of a cluster of intact coffee cells, with a layer of partially broken coffee cells at its surface. Within the particle, many pores, or small holes in the cell walls, connect the intact coffee cells to one another. These pores, widened by the roasting process, create a network of paths through which water will be able to enter. When water first comes into contact with the surface of coffee particles, it quickly dissolves and washes out (*advects*) the solubles that were contained in the broken coffee cells. These chemicals not only get dissolved in the water, but they also quickly get carried away far from the coffee particle. As this happens, water also penetrates inside the coffee cells through its pores and begins dissolving the chemicals inside the particles.

This, in turn, allows the process of *diffusion* to happen. The washing out of the particle's surface makes the chemical compounds in that region much less concentrated, and therefore, diffusion will slowly bring out some of the chemicals from the inside of the particle (an area of higher concentration) up to the surface. As they reach the surface, these compounds immediately encounter some fast-moving water and, yet again, they get advected away. Therefore, the process of coffee extraction takes place in two distinct steps: a slow process of diffusion that brings chemicals from the core to the surface of the coffee particles, and a fast process of advection once they reach the surface.

1.3 Particle Size

Some coffee particles, sometimes called *fines*, are so small that they consist of only broken coffee cells. As you can imagine, those quickly give away everything that is soluble as they get washed out by the flow of water. Because they give away all their solubles at once, the contribution of fines distinctively affects the flavor profile of coffee. Some chemical compounds are much slower than others at diffusion, and because of that they remain entrapped in larger coffee particles. Coffee oils, for example, tend to remain in larger particles;[1] other compounds, such as trigonelline and caffeine, diffuse very quickly, and they will come out, even out of the larger coffee particles, without too much trouble (Moroney et al. 2019). Because of these varying diffusion rates of different chemical species, *changing the grind size will have a strong impact on the flavor profile of coffee, even if brew time is adjusted to obtain the same average extraction yield.* Human taste is a highly nonlinear phenomenon (Illy and Viani 2004; Moroney et al. 2019), and small changes in chemical composition can have an outsized effect on flavor perception.

Brewing ground coffee that contains particles of different sizes will affect the physics of percolation. While this last point may be intuitive, it is important to keep in mind that changing the grind size will also inevitably alter the flavor profile. In other words, even if we mastered the art of percolation to the point of obtaining a perfectly even flow of water through the coffee bed, we may still have a strong preference for a specific grind size when brewing filter coffee.

1.4 Temperature

In the previous section, we learned that before a specific chemical compound can contribute to the flavor profile of your coffee, it must first be dissolved by water and diffuse toward the surface of the coffee particle, where the flow of water will then carry it to the cup of coffee. Those first two steps happen at different rates for different chemical compounds, and they are also affected significantly by the temperature of the water.

Let's imagine an initially dry, solid mass of flavor compounds located in a coffee cell, well below the surface of a coffee particle. When it encounters fresh water, the mass will dissolve initially very quickly, and its rate of dissolution will slow down as the surrounding water becomes more concentrated. This is expressed by the Noyes–Whitney equation (Smith 2015):

$$\frac{\mathrm{d}m}{\mathrm{d}t} = A\frac{D}{L}\left(C_{\mathrm{sat}} - C_{\mathrm{sol}}\right)$$

In the equation above, $\mathrm{d}m/\mathrm{d}t$ at the left represents the rate at which material is being removed from the dry mass as the compounds dissolve. A is the

[1] even if, technically, oils do not dissolve in water

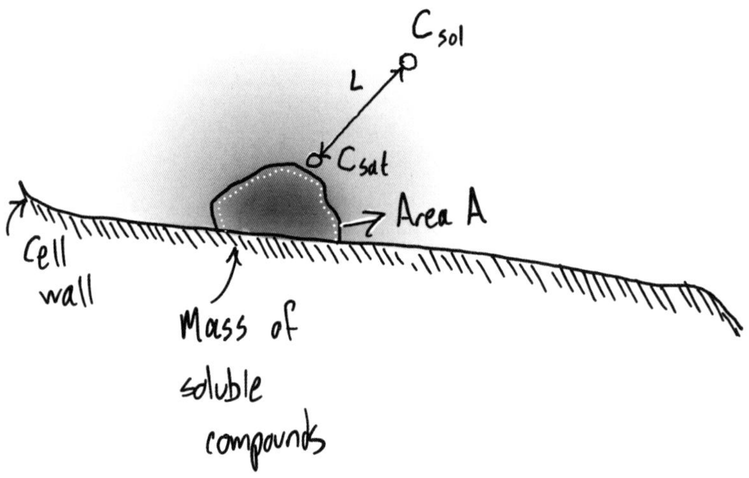

A schematic illustrating the terms used in the Noyes–Whitney equation

total surface area of the dry mass of solubles in contact with water, and C_{sat} is the concentration of water very close to the mass (also called the saturation concentration). C_{sol} is the concentration farther away from the mass, at a distance L. Overall, the term between parentheses represents how fast the concentration drops in the neighborhood of the mass of dry solubles. It represents the fact that the rate of dissolution is faster when the surrounding water is cleaner. As you can imagine, if something were to agitate water inside the coffee cell (which would be very hard to do in practice), the distance L at which the concentration drops by the same amount would be reduced for a moment, and the rate of dissolution would increase.

One of the more interesting variables in the Noyes–Whitney equation is D, the diffusion coefficient. As you can see in the previous equation, a larger value for D will increase the rate of dissolution. This variable D is the quantity that varies across chemical compounds, as dictated by the Einstein–Smoluchowski relation (Lindsay 2009):

$$D = \mu_{\mathrm{p}} k_{\mathrm{B}} T$$

This equation simply states that the diffusion coefficient of a chemical is the product of three things: its mobility μ_{p}, which is highly specific to each chemical compound and is related to how fast it can travel in a given fluid, Boltzmann's constant k_{B},[2] and the temperature T. In other words, the warmer the water, the faster it will dissolve masses of dry chemicals inside the coffee cells.

But things do not end there. The diffusion coefficient above also represents how fast the dissolved chemical compounds will diffuse toward the surface of a

[2] Boltzmann's constant is a fundamental constant of physics. Its value is always 1.381×10^{-23} joules per kelvin.

particle of ground coffee. Therefore, hotter water will not only dissolve chemicals faster, but it will also bring them to the surface faster. In the absence of obstacles or flow, the solution of the diffusion equation tells us how the concentration c varies around a region of higher concentration c_0:

$$c\,(r, t) = c_0 \exp\left(-\frac{r^2}{4Dt}\right)$$

In the equation above, r represents the distance from the center of the more concentrated region, t is time since the diffusion started, and D is yet again the diffusion coefficient. It is probably easier to interpret this equation with a figure.

The graph below shows the concentration profile along one direction, at different moments in time t. It follows the form of a normal distribution, also often called a bell curve, and its shape becomes wider as time passes. It will eventually lead to a uniform concentration, just as the diffusion process does. This concept suddenly becomes a lot more interesting if we look at the concentration profiles of different chemical compounds at a fixed time t, with their own different mobilities μ_p that will cause them to have different diffusion coefficients D, as shown on the following page. I placed a vertical line where we could imagine the surface of a coffee particle to be located. If we look at the concentration of different compounds at the surface of the particle, we get an idea of the rate at

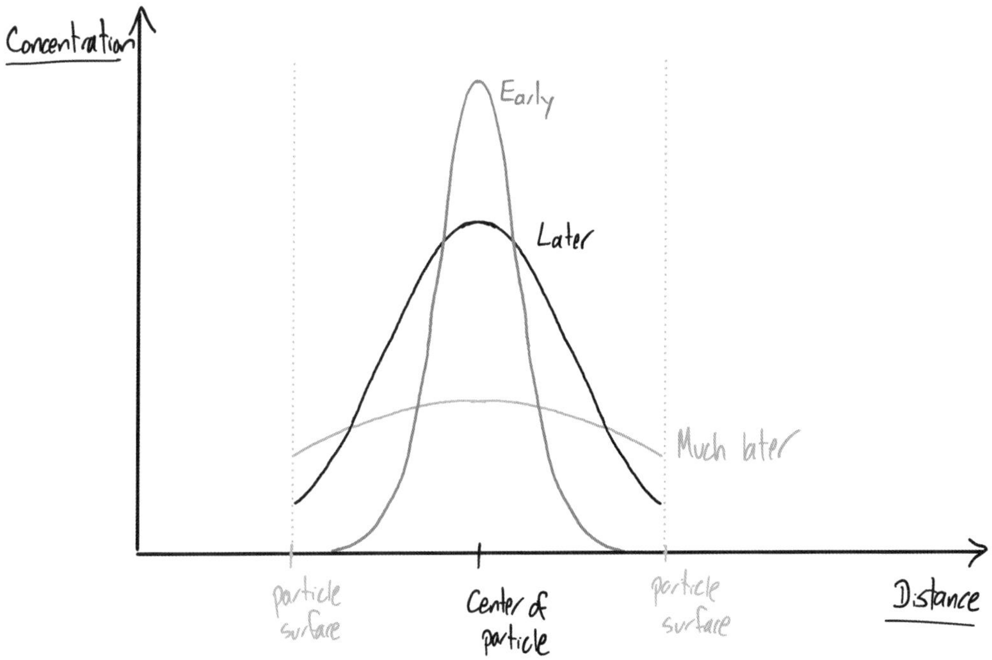

Profile of concentration for a single coffee chemical diffusing to the particle surface over time

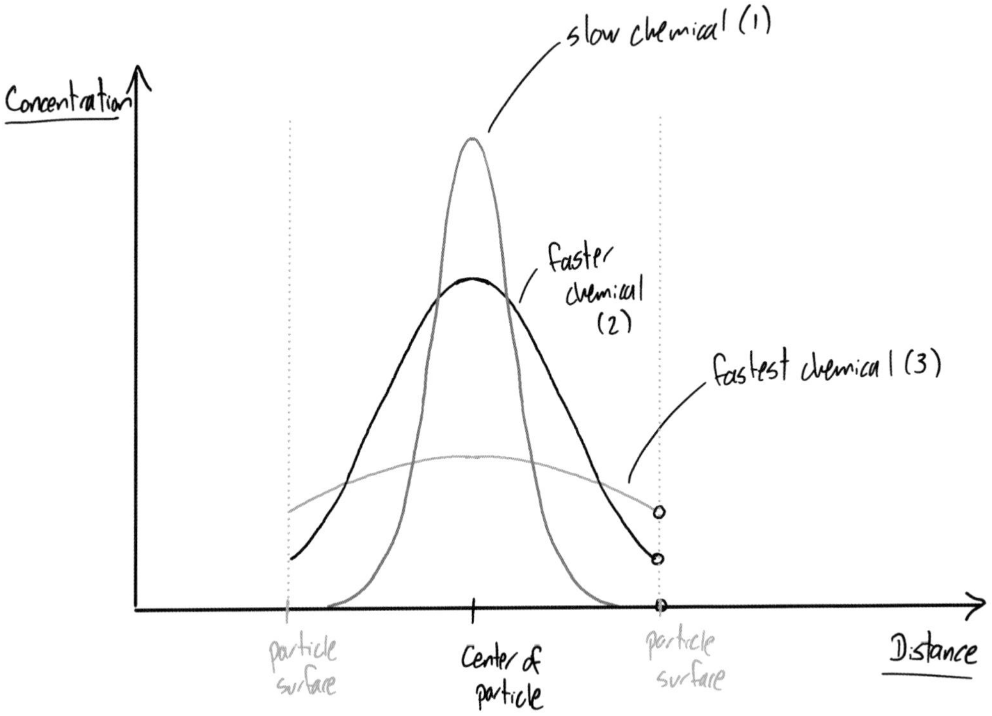

Profile of concentration for different coffee chemicals diffusing to the particle surface at a fixed moment in time

which these chemicals are "leaking" from the coffee particle at this moment in time.

In real life, many obstacles (e.g., cell walls and their pores) will also affect the concentration profile, but they will not change this conceptual picture. The collective effect of the coffee cells can be accounted for with a modified diffusion coefficient dictated by the properties of the coffee particle as a porous medium. Simply put, the diffusion coefficients of all chemicals get drastically reduced, as one might have expected (Grathwohl 1998). Something interesting happens if we make water much warmer; the diffusion coefficients of all chemical species get larger, and the concentration profiles widen accordingly. If we take another look at the concentration values at the surface of the coffee particles, we get a very different profile of extraction rates, shown opposite.

Changing the temperature T will not only change how fast extraction happens, but it will also change the profile of chemicals that end up in the cup of coffee. Changing the size of a coffee particle has an effect very similar to that of changing the water temperature, as shown in the last two figures: instead of changing the width of the concentration profiles, though, we are changing the location of the vertical line that represents the surface of the coffee particle. It could therefore be expected that either grinding finer or using higher brew temperatures would affect the flavor profile in a similar way.

However, other considerations will make these effects of brew temperature and grind size distinct. First, we will see in Chapter 4 that changing the water temperature or the average coffee particle sizes has distinct effects on the physics of percolation (specifically, on how water flows through the coffee bed). We will also see, in Chapter 3, that changing the grind size not only changes the average size of coffee particles; it also drastically changes the fraction of fine coffee particles that are made up of broken coffee cells. As already mentioned, these fine particles will quickly give away all soluble chemicals they contain, which causes a distinct effect on the taste profile, different from that of changing brew temperature. Thus, temperature and grind size will both have specific and unique impacts on the taste profile of coffee, even if they bear some partial resemblance.

In my experience, playing with brew temperature is most useful when I am dealing with darker-roasted coffee; I tend to prefer using water off-boil for light roasts when brewing in a *V60,* and I lower that temperature to anywhere between 88°C and 96°C (190–205°F) for darker-roasted coffee, depending on how dark it is. Some preliminary work demonstrated that *slurry* temperatures above 90°C (194°F) contain more sugary compounds than 74°C (165°F) extractions, in the case of espresso (Easthope 2015). This is consistent with my experience when brewing lightly roasted filter coffee; using cooler brew temperatures reduces the

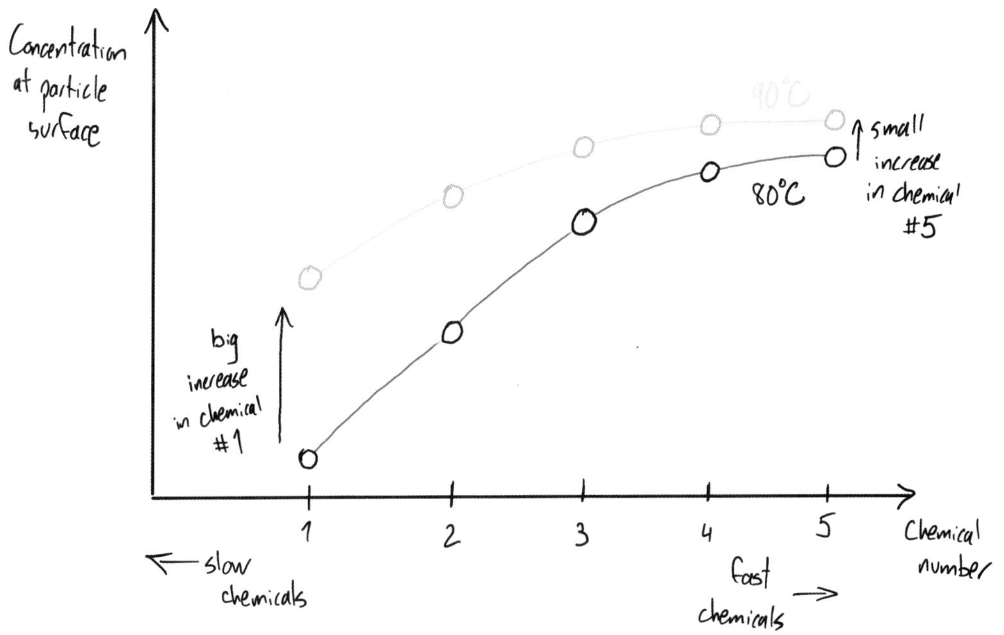

Profile of chemical concentration at the surface of coffee particles for two distinct temperatures

sensation of sweetness. I will discuss this further in Chapter 10, but we will also see, in Chapter 7, that the same kettle temperature can result in different slurry temperatures for different drippers.

1.5 Immersion and Percolation

In the previous section, we discussed considerations of extraction rates in the context of percolation, because we assumed particles of coffee were placed within a flow of clean water. Things are different in the case of an *immersion,* where water and coffee particles remain together and at rest: in that context, there is typically no flow of water outside the coffee particles unless the barista actively agitates the slurry. In the absence of flow, only diffusion can propagate the chemical solubles away from the surface of coffee particles and farther into the slurry. Therefore, the much slower diffusion process will be responsible for carrying the chemicals from inside the coffee particles to their surface and onward.

As an immersion progresses, the concentration of compounds in the slurry water continuously increases. As we saw in the last section by inspecting the Noyes–Whitney equation, the rate of extraction will slow down with time. This makes immersion brews less sensitive to brew time and more easily repeatable, because the later parts of the brewing process extract more gently. However, they are not as powerful at extracting efficiently, and they will also result in cups of coffee that have more *body.* This latter part is explained by coffee fines and tiny particles in suspension making their way into the beverage. Only percolation can typically achieve a very clean beverage, free of undissolved particles, because the bed of coffee is great at self-filtration. We will come back to the concept of self-filtration in Chapter 5.

Immersion and percolation brews taste different for yet another reason, and it is related to what happens outside of the coffee particles. In both cases, diffusion brings compounds from inside the coffee particles to their surface. In an immersion brew that is not agitated, diffusion is also responsible for spreading these chemicals into the beverage and controls the rate at which they leave the surface of coffee particles. As the slurry becomes saturated with each coffee compound, their respective extraction rates gradually slow down.[3] In a percolation brew, clean water is constantly flowing around the particles, and it carries compounds with it very efficiently through advection. During a percolation, the

[3] In the context of coffee, the Noyes–Whitney equation can be applied independently for each chemical species; this is true only when the different chemicals are not reacting with each other.

profile of chemicals extracted into the beverage at any point in time will therefore resemble that at the surface of coffee particles, whereas in an immersion, the compounds that diffuse faster will tend to be overrepresented in the beverage.

1.6 Liquid Retained Ratio

The *liquid retained ratio* is a concept that is important to understand for calculating the average extraction yield of a coffee brew. Naively, one could expect the coffee bed to weigh slightly less after the brewing process, because a fraction of its mass was extracted into the cup of coffee. However, another much more important effect determines the weight of a coffee bed after brewing: the coffee grounds retain a significant amount of water—typically, more than twice their dry weight. This is in part due to intact coffee cells absorbing some water, but also to water remaining trapped between the coffee particles due to its *surface tension*.

The phenomenon of surface tension results from water's tendency to minimize its surface that is exposed to air; surface tension is responsible for the formation of drops, and for water remaining trapped in a small enough enclosure such as a narrow tube or a dropper if no pressure difference large enough forces it to escape. The small spacings between particles of coffee act in a way similar to a dropper; some water will remain trapped there unless something forces it out.

The liquid retained ratio (sometimes called LRR) is usually defined as the

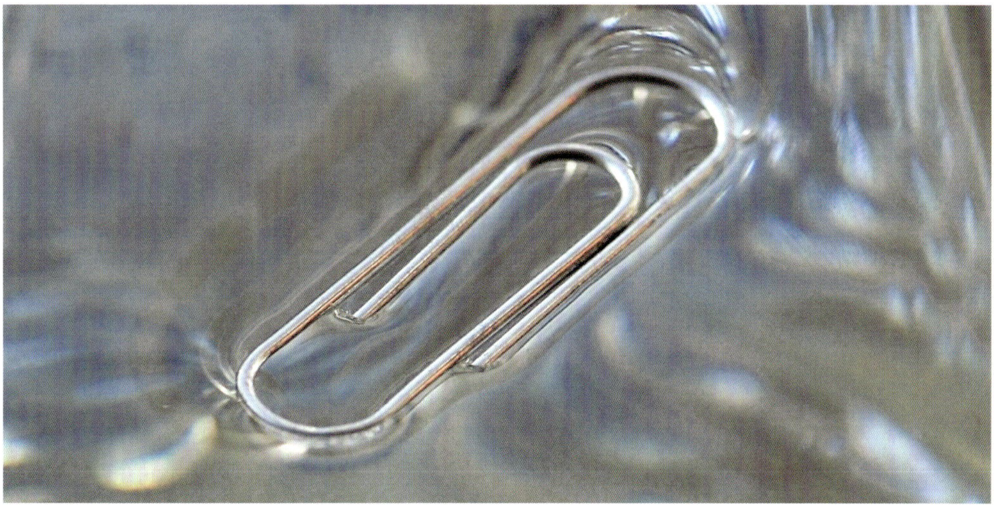

A paper clip floating on the surface of water as a result of surface tension. Source: Wikimedia Commons

total weight of water retained by the coffee bed, divided by the weight of dry coffee before brewing. As a first approximation, the following equation can therefore be used to calculate the liquid retained ratio:

$$\mathrm{LRR} \approx \frac{W - B}{D}$$

where W is the *water weight,* or the total weight of water that was poured on the coffee, B is the resulting weight of the coffee beverage, and D is the weight of the dry *coffee dose.* This equation already outlines that the *beverage weight* is *not* the same as the weight of water poured on the coffee, but it neglects the contribution of dissolved coffee solids to the weight of the beverage. A more accurate expression for the liquid retained ratio is

$$\mathrm{LRR} = \frac{W - B \cdot (1 - C)}{D}$$

where C is the concentration of the coffee beverage, expressed in units between zero and 1. For example, a 1.40% *TDS* coffee would have $C = 1.40/100 = 0.014$. This more accurate expression modifies B so that we consider only the part of the beverage weight that is made of water. For typical V60 beverages (a 1:17 brew ratio with LRR of approximately 2.2), the approximate equation for LRR we gave earlier would underestimate its value by the significant margin of 10%.

1.7 Calculating Extraction Yield

There can be more than one reason to calculate the average extraction yield of a cup of coffee. One could do it simply to compare methods with other people, to understand the impact of changing a variable on the dynamics of brewing, to estimate the fraction of coffee that was wasted, or to attempt to track the taste profile of a cup of coffee. In most of these scenarios, the simplest way to calculate extraction yield is the most useful:

$$\mathrm{EY} = \frac{CB}{D}$$

Namely, extraction yield EY is the product of coffee concentration C with the weight of the resulting coffee beverage B, divided by the weight of the dry coffee dose D. We will refer to the equation above as the *pure percolation case.* As mentioned in the previous section, the weight of the coffee beverage B is different from the weight of water that was poured on the coffee because it excludes some water that remained in the spent coffee bed, and it includes a small amount of chemicals that were extracted from the dry coffee grounds. I recommend using this simple equation when communicating measurements of extraction yield with other people. Its simplicity decreases the chances that different baristas are using different assumptions, and it minimizes the chances of making a calculation mistake. In a context where a barista is trying to estimate the amount of coffee solubles that were consumed versus wasted, the simple equation above will provide the best measurements, so it should also be used.

If a barista is trying to calculate an average extraction yield for the sole purpose of tracking the flavor profile, however, different mathematical forms of the equation should be used to improve how well extraction yield correlates with the chemical profile. The main reason why different equations can be useful revolves around what happens with the water that gets retained in the spent coffee bed. As you might recall from our discussion above, the rates of extraction of different chemical profiles change with time as diffusion transports them from the core to the surface of coffee particles. As time goes on during coffee brewing, the concentration profiles inside coffee particles will get wider, and some coffee compounds may get depleted before others.

Because of this, the quantity that would best track the final flavor profile of the resulting coffee is the final concentration of solubles not only in the cup of coffee, but also in the water that remains trapped between the spent coffee particles. Yes, you read that correctly: even water trapped between the coffee particles at the end of the brew is of interest, because it informs us of the past extraction rate of the brew. It is merely a tool to reconstruct the history of how coffee particles extracted, rather than a direct measurement of what ended up in our cup. By adding up these two contributions, extraction yield will get just a bit closer to a quantity that best correlates with the average profile of chemicals that made their way to the surface of coffee particles during the brew. This will be especially important if we want to compare two brews for which the amount of retained water is very different.

One stark example of a situation where we must consider the retained water would be an immersion where we just scoop out half of the slurry while avoiding the ground coffee. In that situation, it becomes more obvious that calculating the extraction yield based only on the final (scooped out) beverage weight would bias our calculation to an average extraction yield that is far too low, leading us to expect that the brew would taste bad because the coffee was underextracted, while it does not.

One of the difficulties in calculating this more precise extraction yield is that we must factor in the water retained between the coffee particles at the end of the brew, but not that which was absorbed *into* intact cells in coffee particles. In the case of immersion brews, the calculation is much easier because for any reasonable brew time we can approximate that the concentration of chemical compounds in water trapped between coffee particles at the end of the brew will be similar to that of the final coffee beverage. In that case, the equation for average extraction yield becomes

$$\mathrm{EY} = \frac{C}{1-C}\left(\frac{W}{D} - f_{\mathrm{abs}}\right)$$

In the equation above, C is still the coffee concentration, but this time we use W, the total weight of water added to the grounds of coffee, rather than B as we did before. It is important to provide the coffee concentration in fractional form. In other words, C must be a value between zero and 1 when it is inserted in the equation above. If your concentration is 1.4%, for example, enter $C = 1.4/100 = 0.014$. The details of how this equation can be derived are provided in the Appendix. The parameter called f_{abs} is the absorbed water ratio—that is, the weight of water absorbed by the coffee particles divided by the weight of the dry coffee dose before brewing. This ratio is generally between 1.0 and 1.6, but it depends on grinders and grind size. It can be estimated with an *AeroPress* if you let the brew steep for about the same time as for your preferred method and then press as much water as possible out of the dose after brewing. You will then be able to estimate f_{abs} using the equation

$$f_{\mathrm{abs}} \approx \frac{W - B}{D}$$

In the case of a percolation, things get a bit more complicated because the concentration of water retained in the slurry at the end of the brew (C_{slurry}) is usually different from that of the beverage (C_{beverage}). In my experience, V60 brews prepared with a 3:30 brew time using the Weber Workshops EG-1 v2 grinder and SSP ultra-low-fines burrs have a C_{slurry} that is about 70% the value of C_{beverage}, but in general you will need to measure C_{slurry} by sampling the last few drops of your brew and measuring their concentration independently of your beverage concentration, because how C_{slurry} relates to C_{beverage} will likely be dependent on your tools and your brew recipe. The resulting equation for the general percolation case is

$$\mathrm{EY} = \frac{C_{\mathrm{beverage}} - C_{\mathrm{slurry}}}{1 - C_{\mathrm{slurry}}} \cdot \frac{B}{D} + \frac{C_{\mathrm{slurry}}}{1 - C_{\mathrm{slurry}}}\left(\frac{W}{D} - f_{\mathrm{abs}}\right)$$

where variables B, D, W and f_{abs} are the same as described above. As for the immersion equation we described earlier, it is important that $C_{beverage}$ and C_{slurry} are entered as values strictly between zero and 1, rather than as percent values. I do not recommend using the equation above for sharing average extraction yield values online. It should instead be used to compare the characteristics of brews in a way that correlates as much as possible with the taste profile, especially across scenarios where the amount of water retained in the coffee bed is very different.

One scenario where I personally found the previous equation useful was in comparing AeroPress and V60 brews. If an AeroPress brew is prepared by pushing the piston all the way down, very little water remains trapped between the coffee grounds because the pressure pushed it into the beverage, and in that scenario the general extraction yield equation becomes equivalent to the pure percolation case discussed at the beginning of this section. This is contrary to the popular belief that the appropriate equation for AeroPress is the pure immersion case; the confusion arises because AeroPress has an immersion step, but the immersion equation does not hold because the percolation step usually gets rid of any water retained between the coffee grounds. The V60 brew requires the general percolation equation because some concentrated water gets retained between the spent coffee grounds.

Correctly comparing the two methods with the appropriate equations is crucial if a barista wants to determine which method allows one to push average extraction yield as high as possible without obtaining unpleasant flavors, for example. The correct comparison was enlightening for me, as I was initially estimating my AeroPress extraction yields to be much higher than those of my V60 brews, using the immersion equation with the former, and still obtaining flavors typical of underextraction—to my confusion. This also made no sense to me because I was using less water, and immersion extracts more gently than percolation. Using the appropriate equations revealed that my AeroPress brews were indeed underextracted and that grinding finer and using less coffee or more water were the general directions in which I needed to go to improve my recipe.

CHAPTER 2

Water

PHOTO BY KATHY GAGNÉ

Often, when I read discussions about the importance of water in coffee brewing, I encounter an argument that I believe completely misses the point. We are often told, "Water is more than 98% of your cup of coffee, so it better taste good if you want your coffee to taste good" and . . . sure! But this is the part I find *least* interesting about how water affects the taste of coffee. I want to be clear: your water should obviously taste and smell good if you want to make good coffee with it. The part I find much more interesting is that perfectly good-tasting water can yield a very bad or boring cup of coffee. This is something that has only recently been widely appreciated by the coffee community. In this chapter, we explore why this might be the case and give the reader general recommendations for crafting and diagnosing water for coffee.

As this book will review only the basics of water properties for coffee brewing, I recommend *Water for Coffee* by Colonna-Dashwood and Hendon (2015) for further reading on this topic. The Coffee Excellence Center of the Zurich Institute is also actively researching this topic, and I expect interesting new papers to come out on this subject in the upcoming years. The Specialty Coffee Association has also posted some useful documents relating to water properties on their website.

2.1 Water and Dissolution

Before we delve into water recommendations, it is useful to understand the fundamental principle of *dissolution*. Water is a compound with very special properties, and it is no coincidence that it is a central requirement for life as we know it. Water is a molecule composed of two hydrogen atoms (H) bound with an oxygen atom (O) arranged in a triangle shape.

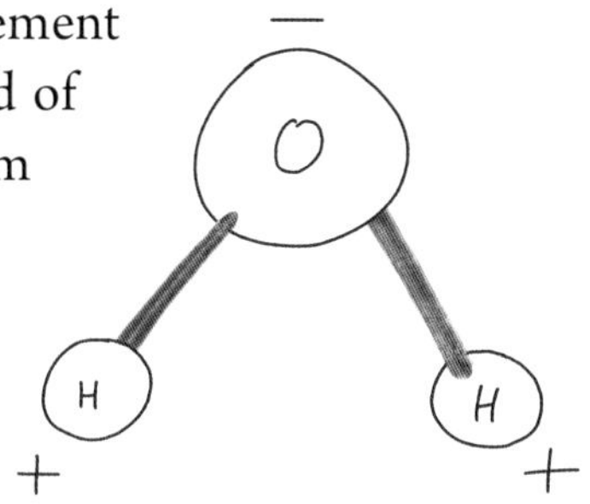

A water molecule with its dipolar charges

The individual atoms are held together by the electromagnetic force, one of the four fundamental forces known to govern our universe. Each atom is composed of a nucleus, which is itself made of protons and neutrons tightly held together with another fundamental force, called the *strong nuclear force,* and electrons floating around the nucleus. Protons carry a positive electric charge, and electrons carry a negative electric charge. As a result, an attractive force between the nucleus and its "cloud" of electrons ensures that the latter do not just fly away. In the stable states of atoms (where we typically find them in everyday life conditions), the total electric charge of the nucleus perfectly balances that of its electron cloud; viewed from a far enough distance, the atom therefore appears to have a neutral charge and does not react strongly to electromagnetic forces.

Some interesting things can happen, however, if two atoms get close enough to each other in the right conditions. One atom can completely steal an electron from another, bringing the total electric charges of both atoms out of balance and causing a strong attraction that will tie the atoms together. This type of *ionic bond* is responsible, for example, for how the sodium (Na) and chlorine (Cl) atoms form solid salt (NaCl). In other cases, one atom will only pull away one or more electrons from another atom without completely stealing it. This partial theft forms another type of bond, called *covalent,* and it is responsible for keeping molecules of water together. In this particular case, the oxygen atom partially pulls an electron away from each of the two hydrogen atoms, and as a result these partially stolen electrons will spend more time closer to the oxygen than to the hydrogen. On average, the oxygen atom will therefore have a slightly more negative charge, and the two hydrogens will have a slightly more positive charge: this is called a dipole moment, and it is why we describe water as a polar molecule.

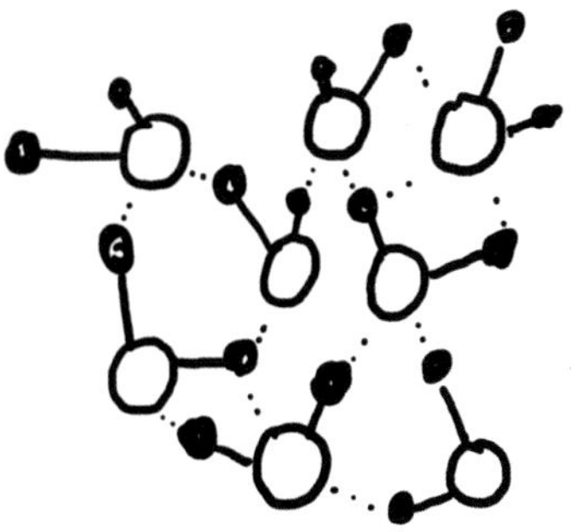
A network of water molecules forming hydrogen bonds

Because of water's polarity, its molecules tend to arrange in a pattern that places the negative sides of one molecule closer to the positive side of its neighbors, as if each molecule were a tiny, weirdly shaped magnet. These forces of attraction between the molecules of water are called hydrogen bonds. The magic of dissolution happens when a foreign molecule is introduced into this network of neatly arranged water molecules. The water molecules quickly react by rearranging themselves around the intruder, presenting their more negative sides to where the intruder is charged positively, and vice versa. The resulting electric attraction can be strong enough to pull apart the new molecule into its individual atoms. As an example, the ionic bond of the NaCl molecule can be broken by this attraction, and Na will get completely separated from Cl.

The story does not end there, however, because Cl leaves with its extra electron it stole from Na, and therefore both the charges of Na and Cl are out of balance (Cl is negative, and Na is positive). For this reason, we call Na and Cl ions. More specifically, the positive ion is called a cation and the negative ion is called an anion. You will also often encounter a small positive or negative sign next to an ion, such as Na^+ or Cl^-, indicating that they are electrically out of balance. In other cases, where ions are out of balance by two or more electrons, you will encounter a number next to the charge sign, such as happens with the magnesium ion Mg^{2+}. Because the ions are out of balance in terms of their electric charges, the surrounding water molecules will keep rearranging themselves to bring back global electrical neutrality as much as possible.

This surrounding of the ions by a shell of properly configured water molecules is what causes the phenomenon of dissolution, and it corresponds to the point where the dissolved solid becomes invisible to our eye. In almost all scenarios where we encounter water, we are in the presence not only of pure H_2O molecules, but also of a fraction of ions that were dissolved into the water at some point in the past. This can happen when rainwater flows through minerals in the ground and partly dissolves them. One typical example of such a mineral is calcium carbonate ($CaCO_3$), which provides water with both bicarbonate ions HCO_3^- and calcium ions Ca^{2+}.

The concentration of bicarbonate ions will affect what we call the *total alkalinity* of water, and the calcium ions will affect its *total hardness.* These are the most important water properties for coffee extraction, *even at concentrations low enough that they do not impact the taste of water by itself.* The total alkalinity tells us how much chemical compounds

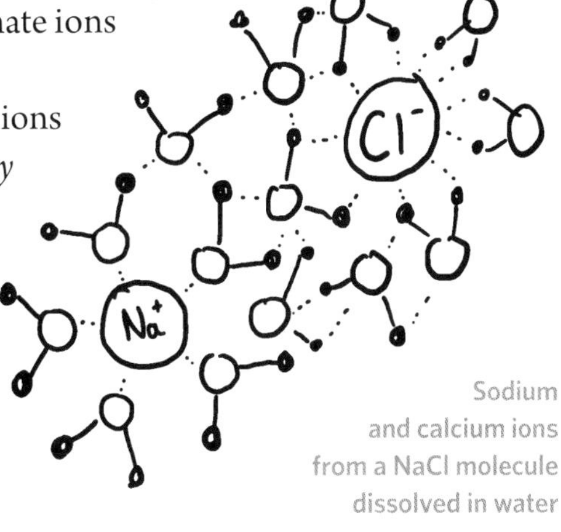

Sodium and calcium ions from a NaCl molecule dissolved in water

	German degree	mmol/L Ca^{2+}+Mg^{2+}	mmol/L HCO_3^-	mg/L Ca^{2+}	mg/L Mg^{2+}	mg/L HCO_3^-
1 ppm as $CaCO_3$ =	0.056	0.01	0.02	0.40	0.24	1.22

in the water will fight against changes in pH whenever acids are added.[4] This is crucial because it means water with a high total alkalinity will *denature some good-tasting acids* that extract from the coffee beans before we can taste them. The second important property is the total hardness of water, or its concentration in positive ions, which is believed to play a role in extracting the flavor compounds from the coffee beans. We discuss these two properties in more detail below.

The concentration of ions can be measured in several different units. One system often used is based on mass concentrations, where the weight of ions is divided by the volume of water in which it was diluted, such as milligrams per liter (mg/L). This unit is also sometimes called parts per million (ppm), especially on American bottled waters. I personally find this misleading because it could be interpreted as a number of ions per million total molecules, which ppm does *not* stand for. Mass concentrations are impractical for several reasons; chemists often prefer to *count* the number of ions because this makes it easier to account for the effect of chemical reactions on concentrations.

There is, therefore, an alternative system of so-called "equivalent concentrations" that scale with the number of ions, rather than their weights. This is important because single ions of different chemical species (such as Ca^{2+} or Mg^{2+}) have different weights. Examples of "equivalent units" include moles per liter (mol/L), where a mole is just a fancy word for a large number of molecules (6.022×10^{23} molecules). Other units that fall in this category include "ppm as $CaCO_3$." One ppm as $CaCO_3$ represents the concentration by weight that you would obtain if you replaced your ion with a molecule of $CaCO_3$. It is a weird way to obtain a concentration that scales with the *number of ions*, rather than their weight. These units behave as if the ion in question had the molecular mass of a $CaCO_3$ molecule, and thus this system ensures that the measurement does not depend on the molecular mass of the chemical species in which you are interested. This is the standard unit adopted by the Specialty Coffee Association (e.g., Wellinger, Smrke, and Yeretzian 2018), and we will therefore use it throughout this book. The table shown above will allow you to convert 1 ppm as $CaCO_3$ to different units of concentration.

4 pH, or hydrogen potential, is a property that determines whether a solution is acidic (low pH) or alkaline (high pH). An "alkaline" pH is a different concept from that of "total alkalinity," despite their similar names.

2.2 Total Alkalinity

The *total alkalinity* is probably the most important parameter that will affect the taste of coffee, apart from contaminants that could impart a bad taste to the water itself. In most scenarios,[5] the most important variable that determines the total alkalinity is the concentration in HCO_3^- ions. The total alkalinity of water represents its ability to resist changes in pH. If acids are added to the water, they will first get modified by reacting chemically with HCO_3^- ions until those get depleted. Only then will adding more acids dramatically shift the pH of water. A lot of chemical compounds that are responsible for the desirable taste of coffee are acids, and it is therefore not a good thing to have a high total alkalinity, because *acids will get modified by reacting chemically with* HCO_3^- *ions as they extract*. It is likely that some small level of total alkalinity is desirable for coffee preparation, but preliminary results based on sensory analyses show that water having the lowest total alkalinity among those tested, 20 ppm as $CaCO_3$, was preferred by tasters (Wellinger and Yeretzian 2015). A total alkalinity in the range of 40–50 ppm as $CaCO_3$ is more commonly used in the specialty coffee community, and I personally find that 20 ppm as $CaCO_3$ or lower produces brews that are a bit too sour for my preference with light-roasted coffee. Note that, although this is not specified in the Wellinger and Yeretzian (2015) study, the coffee that the researchers used may have been roasted darker than is typical for average specialty coffee. If this is the case, it is likely that the tasters favored a lower total alkalinity than they would have when drinking lighter-roasted coffee. Using a lower total alkalinity in this context could have preserved more perceived acidity, which tends to be otherwise less prominent in darker roasts.

2.3 Total Hardness

The concentration in positively charged ions determines another important property of water, namely its total hardness. Its effect on coffee taste is not as straightforward as that of total alkalinity, but the consensus currently accepted by most coffee professionals is that *harder water* favors the extraction of some specific chemical compounds. The exact mechanism to explain how positively charged ions affect extraction has not been demonstrated clearly in the context of coffee, but researchers hypothesize that such ions can increase the diffusion rate of specific chemicals, based on gas chromatography experiments (Colonna-Dashwood and Hendon 2015).

[5] when the pH of water is below 8.3 (Baird, Eaton, and Rice 2012)

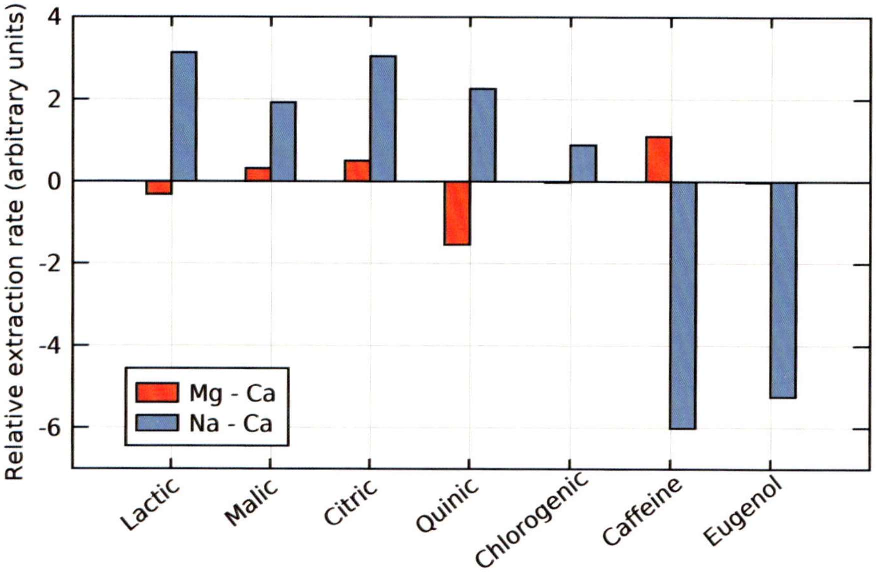

Relative expected extraction rates in the presence of magnesium (red) or sodium (blue) compared with calcium, based on gas chromatography data. Magnesium would tend to extract slightly more caffeine and less quinic acid compared with calcium, whereas sodium would favor a very different profile where several acids are favored and caffeine and eugenol are disfavored. Note that the effect of sodium requires larger concentrations to be apparent. Sources: Barista Hustle and Christopher Hendon

Other preliminary experiments could not successfully measure a difference in extraction rates when varying the total hardness of brew water (Smrke, priv. comm.) and instead hypothesize that calcium and magnesium may act as taste modulators even at concentrations too weak to be perceived by tasters in the water itself. Anecdotal evidence suggests that a higher calcium hardness favors perception of body and sweetness but that including some magnesium along with it can positively affect the complexity of the flavor profile. This area clearly needs further research, and we should expect water recommendations and popular brew water recipes to change in the upcoming years. Colonna-Dashwood (2016, Jan. 5) noted that a good approximate rule to keep in mind is that softer water (having both a lower total alkalinity and total hardness) will tend to extract less and produce a brighter acidity, therefore mimicking some aspects of a lighter-roasted coffee.

2.4 Permanent and Temporary Hardness

The concepts of *permanent* and *temporary hardness* are sometimes used instead of total alkalinity and total hardness when describing water properties. They are especially useful in discussions about the formation of scale. Scaling occurs when positively charged ions recombine with negatively charged bicarbonate ions to form $CaCO_3$, which then *precipitates* (forms a solid) and can damage coffee equipment such as espresso machines.

Temporary hardness, also called *carbonate hardness*, is an indication of how much total scale can be formed, and it is therefore equivalent to the quantity that limits the formation of scale, either total alkalinity or total hardness, whichever of the two is smaller when expressed in equivalent concentrations. This is usually the total alkalinity, because it is smaller than total hardness in most everyday scenarios. Commercial *titration* kits used to measure total alkalinity are often mistakenly called carbonate hardness tests.

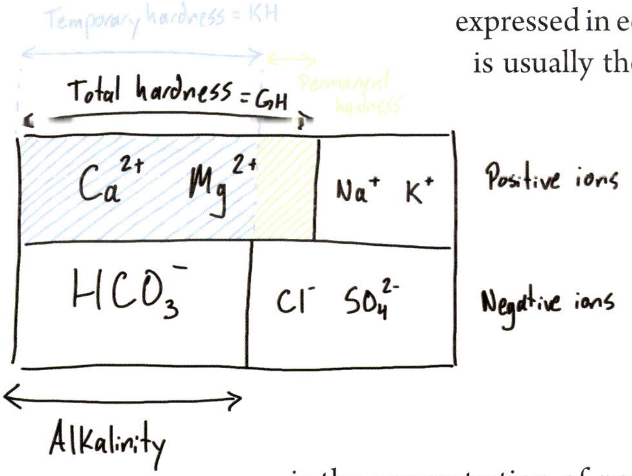

Permanent hardness, or non-carbonate hardness, is the concentration of positively charged ions that will remain in the water even after it formed as much scale as it could, therefore depleting its bicarbonate ions. In most scenarios where total hardness is higher than total alkalinity, we measure it by subtracting the two values expressed in equivalent units. In the more unusual scenario where total alkalinity is higher than total hardness, the permanent hardness is defined as zero.

The chart on the following page shows the total hardness and alkalinity of tap water in different cities (triangles), specific brands of bottled waters (stars), and regions recommended by resources on water for coffee brewing (Colonna-Dashwood and Hendon 2015; Wellinger, Smrke, and Yeretzian 2018). Specific brew water recipes described by various resources (Perger 2017, May 1; Rao 2008; Rao 2013) and the popular Third Wave Water brand of dry minerals are also shown as circles. The dashed line shows the region where total hardness and total alkalinity are in balance, along which most tap waters typically fall. This figure shows that typical recommendations vary, but all are made of relatively *soft water* with a total hardness below 200 ppm as $CaCO_3$ and a total alkalinity in the range 20–80 ppm as $CaCO_3$; most cities across the world have tap water much harder than this.

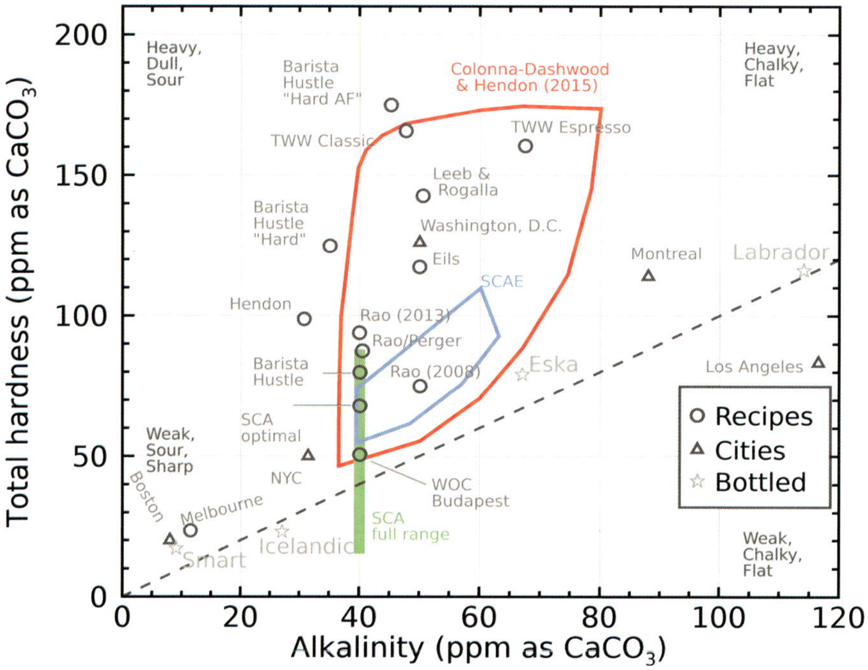

Water compositions recommended by various authors and institutions, compared with recipes, tap water from different cities, and bottled waters

2.5 Ingredients for Crafting Brew Water

Baristas can use several solid-form minerals to craft their own brew water. Historically, the minerals most often used in the coffee community have been the following:

- Calcium chloride ($CaCl_2$)
- Magnesium sulfate ($MgSO_4$)
- Magnesium chloride ($MgCl_2$)
- Sodium bicarbonate, or baking soda ($NaHCO_3$)—*not* baking powder
- Potassium bicarbonate ($KHCO_3$)

but other options include

- Calcium carbonate ($CaCO_3$)
- Calcium sulfate ($CaSO_4$)
- Calcium citrate ($Ca_3(C_6H_5O_7)_2$)

The latter three are either harder to obtain or less practical to use. For example, $CaCO_3$ is much more difficult to dissolve in pure water, and calcium citrate affects the total alkalinity of water through a negatively charged ion ($C_6H_5O_7^-$) different from our usual bicarbonate, complicating the calculations of water properties.

It is very important to choose food-grade minerals when crafting brew water; this requirement ensures some level of purity (depending on the country) and it ensures that the impurities are not harmful to human health. Other options, such as laboratory-grade minerals, may be purer, but they might contain impurities that are unfit for consumption or even poisonous, such as heavy metals.

Another important aspect to pay attention to is the hydrous form of minerals. Hydrous forms, also called hydrates, specify a number of water molecules that are attached to the dry mineral through weak intermolecular bonds, and they will affect the weight of your minerals. If you wrongly assume that your mineral is made of pure $CaCl_2$, for example, unaware that every molecule really comes with four molecules of water, simply weighing your dry minerals will cause you to greatly underestimate the amount of calcium ions that you are really adding to your water. Hydrous forms include the following:

- Calcium chloride anhydrous (no H_2O): most common form, small white spheres
- Calcium chloride hexahydrate (6 H_2O), tetrahydrate (4 H_2O), or monohydrate (1 H_2O): all are less common
- Magnesium sulfate heptahydrate, i.e., Epsom salt ($MgSO_4 \cdot 7\ H_2O$): most common form, small snow-like transparent crystals
- Magnesium sulfate hexahydrate (6 H_2O), pentahydrate (5 H_2O), tetrahydrate (4 H_2O), monohydrate (1 H_2O), anhydrous (no H_2O, i.e., pure $MgSO_4$): all much less common than the heptahydrate form
- Magnesium chloride hexahydrate ($MgCl_2 \cdot 6\ H_2O$): most common form, large crystalline and milky flakes
- Magnesium chloride anhydrous (no H_2O): white powder, less common
- Neither baking soda nor potassium bicarbonate can have hydrous forms

The most important water properties that these different compounds allow us to control are the total alkalinity (through the addition of baking soda or potassium bicarbonate), the total hardness, and the calcium-to-magnesium ratio (through the addition of $CaCl_2$, $MgCl_2$, or $MgSO_4$). Because we have three properties and five ingredients to control them with, we can achieve specific water properties by using different combinations of these ingredients.

For example, the magnesium hardness can be controlled by adding either $MgCl_2$ or $MgSO_4$; the resulting water will have different concentrations in negative ions (Cl^- or SO_4^-), but it is not currently known whether those have an impact on the taste of coffee, and the main reason for using two different minerals is rather historical and based on the availability of the respective minerals rather than being intentional. The distinction between $NaHCO_3$ and $KHCO_3$ for controlling total alkalinity is possibly of some importance because Na^+ may have a small impact on the flavor profile, although much less so than Mg^{2+} and Ca^{2+} (Colonna-Dashwood and Hendon 2015).

2.6 How to Obtain Pure Water

Another important aspect of crafting brew water is obtaining pure water to which dry minerals can be added. This is often the most difficult part of crafting brew water—and the one that will usually be the least environmentally friendly. The available options for obtaining pure water include

- Distillation
- Microfiltration
- Bottled water
- Reverse osmosis

Each has advantages and inconveniences, and frankly, none of these options is amazing. The best possible situation you can be in is probably living in a city where the water is very soft, which can allow you to add mineral concentrates until you reach the desired water properties. Later in this chapter, we explore how to do this.

Distillation can be done at home with some relatively cheap devices available online. They work by heating up tap water until it evaporates and gets collected in a large plastic bottle, leaving the minerals to deposit in the evaporation tank. This is a lengthy process, as it takes about an hour per liter (approximately 1 quart) of water, and usually comes with a noisy fan to avoid overheating the machine. On top of that, expect the apparatus to need regular maintenance because the deposited minerals have to be cleaned out regularly with a weak acid. This option is a lot of work and quite messy. If you don't clean up the tank often enough, you can end up with a weird metallic taste in your water. I used this option for about a month, as I thought this would be a more environmentally friendly option, given that Québec uses mostly hydroelectricity—until my machine overheated despite the fan and broke down. This was definitely not a win for the environment.

Microfiltration pitchers such as the ZeroWater[6] are another solution avail-

[6] This is different from other popular options such as the Brita® pitcher, which is much less aggressive at removing minerals from tap water.

able to craft water for coffee. With these you can remove the vast majority of dissolved solids from tap water and then add back minerals in concentrate form as desired. Peak Water is another recently released option designed by Maxwell Colonna-Dashwood specifically for coffee, where tap water is microfiltered and remineralized directly with ion exchange resins. Another option is the BWT pitcher, which adds back some magnesium to the water after filtration. The downside of these water pitchers is that you will need to regularly filter the water and change the filter often—usually every month or two. If you don't use the pitcher regularly, this will not be a good option because the filter will need to be changed regardless to discourage microbial buildup.

Bottled distilled or microfiltered water can be bought at most convenience stores, and this can be a good short-term or travel option. It can be quite wasteful in terms of plastic bottles, but I suspect the step of microfiltration or distillation is much less wasteful when done in larger batches by the manufacturers. Some companies sell large bottles of water that can be regularly refilled and exchanged; it is worth seeking out this option near you because it may be one of the most efficient and environmentally friendly options listed here. It is important to use plastic bottles made for that purpose because not all plastics are safe for multiple refills or repeated use.

Reverse osmosis systems, which can be installed on your tap water faucet, use the pressure of your incoming water to perform something similar to microfiltration. This option has a higher startup cost, and it will generally give you water that is not quite distilled (most often, you will get 10 ppm water), but this is not a problem because you can easily adapt your mineral concentrates for that. Furthermore, if you like the magnesium-to-calcium ratio and the hardness-to-alkalinity ratio of your tap water, you can adjust a reverse osmosis system such that you won't even need to add minerals back to the water. This is definitely one of the best options for coffee shops and heavy users, but it is also wasteful in a different way: most reverse osmosis systems have an efficiency of 10% to 20%. This means every liter of soft water that comes out of it required letting another 4 to 9 liters of tap water down the drain, even though you won't notice it.

Because I was not really satisfied with any of these options, I also explored the idea of adding acids to tap water in order to decrease the total alkalinity, while leaving the total hardness intact. It turned out that this was also not a good option: weak acids such as vinegar barely affect the total alkalinity; they react away some of the bicarbonate buffer but leave behind another type of buffer. As far as I understand the chemistry, the only way to do this would be to use strong acids, but those can be incredibly dangerous, depending on their concentration—so I chose not to go there, and I don't recommend it.

I provide brew water recipes in Section 2.8. First, let's discuss techniques for preparing a mineral concentrate solution.

2.7 Preparing a Concentrate

Preparing brew water can be done by simply adding the required amount of dry minerals to pure water—but doing so directly is impractical because it requires extreme precision in weighing the dry minerals, and dissolving them will typically require a lot of agitation and waiting. One solution to deal with both of these problems is to first prepare a concentrate, where a larger amount of dry minerals are mixed in with a small amount of hot,[7] pure water. This concentrate can then be stored and used several times. This additional step not only makes the weighing more accurate, but it also helps to dissolve the minerals faster when they are added to the pure water.

It is important to choose your container carefully to store your concentrate. Be sure no parts of the container are made of metal, because it would quickly corrode. For that reason, I recommend using either plastic or glass. A container that is not perfectly airtight is also desirable because the concentrate will usually degas some CO_2 even long after you prepared it, so a permeable container will prevent it from cracking or otherwise breaking. I like to use a 350-mL plastic squirt bottle made for barbecue sauce; it's convenient for storing, shaking, and dosing the concentrate. However, I avoid placing hot liquids in it because the plastic is probably not safe for that; I always let my concentrate cool down in a glass mason jar first.

Here are the steps I like to follow when preparing a mineral concentrate:

1. Using a small glass or plastic spoon (black or transparent is ideal), weigh each mineral, preferably with a 0.01-gram-precision scale, and place the minerals in a glass mason jar large enough for your desired concentrate volume. A 0.1-gram-precision scale could do the job too, but it's not as precise. I like to use either a black plastic Hario coffee spoon[8] or a small, double-walled glass espresso cup as a weighing container. Weigh the minerals in this order, as they get progressively messier: $CaCl_2$, $MgCl_2$, $MgSO_4$, $NaHCO_3$, and then $KHCO_3$. I recommend wearing plastic gloves for this because dry minerals can be irritating to the skin. Be sure to read and follow all safety precautions

[7] Using hot water is not required, but it will speed up the dissolution and CO_2 degassing.

[8] Plastic Hario spoons have a hydrophobic surface, making them more convenient.

on each mineral package, and again, ensure that they are all food grade (*not* laboratory grade).

2. Heat distilled water up to 90°C (194°F) or warmer.
3. Place your mason jar containing the minerals on a larger scale that has enough capacity to weigh the final concentrate, and tare it.
4. The weighing container or spoon that you used for the minerals probably has some residue stuck on its walls. For this reason, I first pour a small amount of hot, distilled water into the weighing container and then pour that liquid into the mason jar. Then, to finish preparing the concentrate, I slowly pour boiling, distilled water into the mason jar until it reaches the desired water weight. Note that the water weight is not exactly the same as the total concentrate weight, because it should exclude the weight of the dry minerals. The concentrate will degas a lot of CO_2 bubbles; this is normal. Depending on your exact recipe, the concentrate may look milky white and thick.
5. Remove the glass jar from the scale without burning yourself, and place some kind of lid loosely on the jar. Do not screw a lid on, because the concentrate will need to degas a lot of CO_2. Let it cool down for a few hours.
6. When the glass jar no longer feels warm to the touch, use a glass or plastic stick to stir it well, and transfer the contents to your final concentrate bottle. Stirring the concentrate just before the transfer is important because most mineral recipes will cause the concentrate to deposit a lot of solid materials. If some solid deposits remain in the glass jar, you can pour back a bit of concentrate, swirl the contents, and pour that liquid back into the concentrate bottle until no solids are left behind.
7. Keep your concentrate bottle in the refrigerator to ensure it remains safe for consumption. Most concentrates will, however, be a hostile environment for bacterial build up because of their high mineral concentration.
8. When you need to prepare a pitcher of brew water, take out the concentrate, shake it thoroughly immediately before use, and weigh the required amount on a 0.1-gram precision scale. I like to use a plastic Hario coffee spoon as the weighing container for this step too. You can place the spoon in a small espresso-sized glass for stability when weighing.[9]
9. Pour the concentrate into the bottle of pure water, shake it well, and pour a bit of water back onto the spoon. Pour this liquid back into the water bottle, thus including any solid minerals that were left on the spoon.

[9] For most concentrate recipes, the volumetric mass of the concentrate is very close to 1.0 g/mL, which allows using a small, 20-mL plastic syringe to conveniently transfer the concentrate to the distilled water.

10. Ideally, wait at least 30 minutes before using your bottle of brew water. Store it in the refrigerator if you are not going to use it within a week or two, or if you began with tap water. Unlike distilled water, tap water contains bacteria, so it is important to always store your brew water in the refrigerator if you prepared it from tap water or a reverse osmosis system. Reverse osmosis may remove most bacteria if the filters are not damaged, but they do not remove all bacteria (Park and Hu 2010), so it is good practice to refrigerate water produced by this method.

Depending on the exact brew recipe you are using, you may observe some solid particles (precipitates) in the brew water for the first several minutes. Ideally, they should later dissolve, but sometimes they do not completely go away. I think this is likely caused by food-grade impurities in some dry minerals, and it does not worry me for the preparation of filter coffee. You might need to rinse your kettle from time to time, but this is good practice anyway. Particles of undissolved minerals will be filtered by your bed of coffee unless you prepare an immersion brew. To get rid of them, you can prepare different concentrates with each mineral individually, with the same concentration as your normal brew water recipe. This will allow you to determine which mineral caused the precipitate and it will allow you to seek a better source from which to buy that specific mineral.

The strength of your mineral concentrate is a matter of personal preference; higher concentrations allow you to be more precise when weighing dry minerals (up to a point), are more hostile to microbial growth, and require a smaller amount of storage space, but they will create more precipitates and may take longer to dissolve fully in pure water. If you have a concentrate that precipitates a lot, it's important to shake it thoroughly just before use; otherwise, your brew water composition will drift over time.

2.8 Some Brew Water Recipes

In this section, we review some brew water recipes that have been used by various actors in the coffee community and how to craft them exactly, with concentrates. As of this date, I have tried only the Barista Hustle, Rao/Perger, and Simplified Rao/Perger recipes, and I usually prefer the latter. I was not able to tell the difference between the Rao/Perger and Simplified Rao/Perger recipes by taste, and the latter is easier to craft and more convenient to use. I tried the Barista Hustle recipe with only a few different bags of coffee, but each time I perceived noticeably less sweetness in the cup, compared with the same coffee prepared with the Rao/Perger recipe. This is consistent with the current belief that calcium plays a role in the perception of sweetness (the Barista Hustle recipe uses only magnesium to adjust for hardness), although I am not aware of a study that demonstrates this based on robust data.

All the recipes listed below were designed for the preparation of 350 mL of concentrate, and they require using 4 grams of concentrate per liter of pure water to obtain brew water.

Name	$MgSO_4 \cdot 7H_2O$ (g)	$MgCl_2 \cdot 6H_2O$ (g)	$CaCl_2$ (g)	$NaHCO_3$ (g)	$KHCO_3$ (g)	Water (g)	KH (ppm as $CaCO_3$)	GH (ppm as $CaCO_3$)
Rao/Perger	8.8	3.5	2.6	3.0	3.5	328.6	40	88
Simplified Rao/Perger	13.0	0	2.6	6.0	0	328.4	40	88
Rao (2013)	0	7.0	5.3	0	7.0	330.7	40	94
Barista Hustle	17.2	0	0	5.9	0	326.9	40	80

2.9 Creating New Brew Water Recipes

The recipes listed above can be prepared in various ways: with different hydrous forms,[10] in different sizes, with different concentrations of mineral concentrates, with soft but not distilled water, or even with different minerals. Doing so requires modifying the exact weights of dry minerals that are used, based on relatively simple considerations of chemistry. I have put together a Web form that will allow you to adapt these recipes.[11] This form could also allow readers to craft entirely new brew water recipes based on the desired outcome properties—for example, in terms of total hardness, total alkalinity, magnesium-to-calcium ratio, sodium ion (Na^+) concentration, and even concentrations of anions such as SO_4^{2-}. Currently, there are no quantitative data on how these latter properties affect taste, but using this Web form could allow readers to design their own experiments and explore the topic themselves.

Another interesting advantage in crafting custom recipes is that they can

[10] Hydrous forms are versions of the minerals with some water impurity, thus requiring us to use more mineral weight to get the same quantity of ions.

[11] https://bit.ly/2Wi92vq

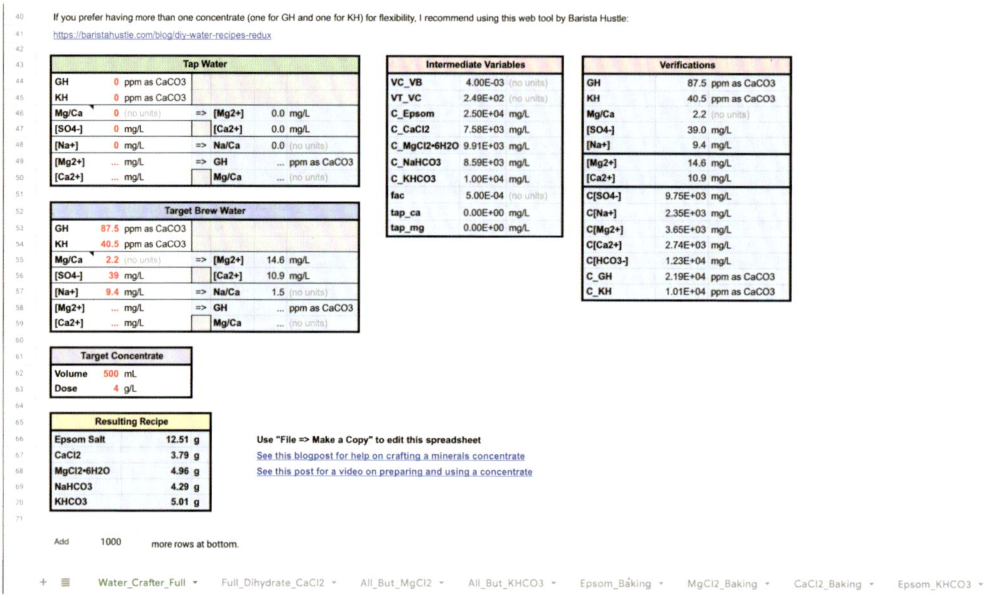

If you prefer having more than one concentrate (one for GH and one for KH) for flexibility, I recommend using this web tool by Barista Hustle:

https://baristahustle.com/blog/diy-water-recipes-redux

Tap Water						
GH	0	ppm as CaCO3				
KH	0	ppm as CaCO3				
Mg/Ca	0	(no units)	=>	[Mg2+]	0.0	mg/L
[SO4-]	0	mg/L		[Ca2+]	0.0	mg/L
[Na+]	0	mg/L	=>	Na/Ca	0.0	(no units)
[Mg2+]	...	mg/L	=>	GH	...	ppm as CaCO3
[Ca2+]	...	mg/L		Mg/Ca	...	(no units)

Intermediate Variables		
VC_VB	4.00E-03	(no units)
VT_VC	2.49E+02	(no units)
C_Epsom	2.50E+04	mg/L
C_CaCl2	7.58E+03	mg/L
C_MgCl2•6H2O	9.91E+03	mg/L
C_NaHCO3	8.59E+03	mg/L
C_KHCO3	1.00E+04	mg/L
fac	5.00E-04	(no units)
tap_ca	0.00E+00	mg/L
tap_mg	0.00E+00	mg/L

Verifications		
GH	87.5	ppm as CaCO3
KH	40.5	ppm as CaCO3
Mg/Ca	2.2	(no units)
[SO4-]	39.0	mg/L
[Na+]	9.4	mg/L
[Mg2+]	14.6	mg/L
[Ca2+]	10.9	mg/L
C[SO4-]	9.75E+03	mg/L
C[Na+]	2.35E+03	mg/L
C[Mg2+]	3.65E+03	mg/L
C[Ca2+]	2.74E+03	mg/L
C[HCO3-]	1.23E+04	mg/L
C_GH	2.19E+04	ppm as CaCO3
C_KH	1.01E+04	ppm as CaCO3

Target Brew Water						
GH	87.5	ppm as CaCO3				
KH	40.5	ppm as CaCO3				
Mg/Ca	2.2	(no units)	=>	[Mg2+]	14.6	mg/L
[SO4-]	39	mg/L		[Ca2+]	10.9	mg/L
[Na+]	9.4	mg/L	=>	Na/Ca	1.5	(no units)
[Mg2+]	...	mg/L	=>	GH	...	ppm as CaCO3
[Ca2+]	...	mg/L		Mg/Ca	...	(no units)

Target Concentrate		
Volume	500	mL
Dose	4	g/L

Resulting Recipe		
Epsom Salt	12.51	g
CaCl2	3.79	g
MgCl2•6H2O	4.96	g
NaHCO3	4.29	g
KHCO3	5.01	g

Use "File => Make a Copy" to edit this spreadsheet

See this blogpost for help on crafting a minerals concentrate

See this post for a video on preparing and using a concentrate

Add 1000 more rows at bottom.

+ Water_Crafter_Full Full_Dihydrate_CaCl2 All_But_MgCl2 All_But_KHCO3 Epsom_Baking MgCl2_Baking CaCl2_Baking Epsom_KHCO3

The water crafter tool allows users to specify their target water properties in a Google spreadsheet and obtain a list of dry mineral ingredients to craft it

be built from tap or reverse osmosis water, assuming that the initial water has a smaller concentration of each ionic species than the target brew water. This is only true because *adding* dry minerals to water cannot decrease its ionic concentrations, unless we purposely caused chemical reactions to filter out some of the minerals. We will not discuss that topic here.

The main difficulty in crafting a brand new brew water recipe to the desired properties is that the barista can affect them only through the respective dry mineral weights used in preparing the concentrate. Some minerals will affect more than one property at once; for example, adding more $MgCl_2$ will increase total hardness as well as change the magnesium-to-calcium ratio. Determining the exact weight that must be used for each mineral to achieve specific water properties requires solving some relatively simple equations. The Web calculator also allows baristas to do this without having to solve any equation.

2.10 Measuring Brew Water Properties

The ability to measure the total hardness and alkalinity of water is important in the preparation of mineral water recipes. The measurements may be used to verify that you prepared the brew water correctly or to determine whether your tap water is suitable for coffee, with or without the addition of dry minerals. Here, we will review some of the methods available to make these measurements.

For a long time, the coffee community has relied on measuring the electric conductivity of water to estimate its total dissolved solids (TDS), via electric conductivity meters. They are convenient because they are small, cheap, and quick to use. They are also problematic: the readings are highly dependent on temperature, and they use a chemical profile of typical tap water to translate the conductivity to a TDS measurement.[12] These devices generally attempt to correct for temperature deviations, but they do so in an imperfect way. Furthermore, the TDS measurement is a rough approximation of total hardness, but it does not inform the barista about the total alkalinity of water, which is an even more important property. Electric conductivity meters are still useful if you want to quickly monitor your tap water or even your crafted brew water, because they can tell you *whether something changed* in your water, albeit with a low accuracy. Typically, they are accurate to about 30% (Wellinger, Smrke, and Yeretzian 2017). They are also useful to verify whether a source of water is distilled or not. I recommend always taking measurements at the same sample temperature. A temperature of 25°C (77°F) is most commonly used to report TDS measurements. Wellinger, Smrke, and Yeretzian (2017) also report that not correcting for the effect of temperature can cause an additional source of error as large as 20%.

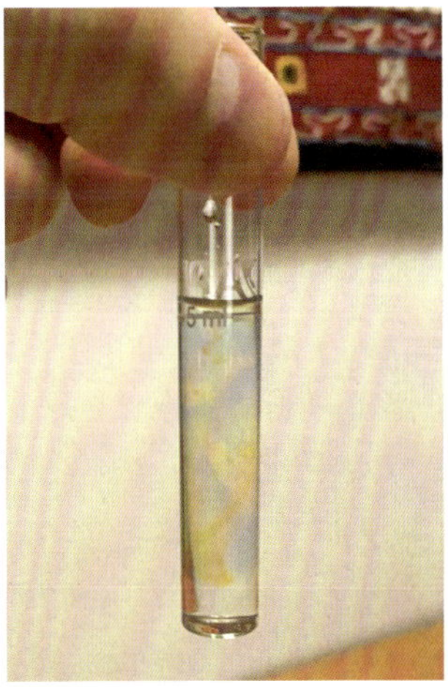

Titration with an API aquarium test kit

Aquarium test kits provide a much better way to measure brew water properties, namely its total alkalinity (often with what is loosely called a KH test kit) and its total hardness (with a GH test kit). I recommend focusing on measuring KH (alkalinity) first, because bad KH can hurt the taste of coffee much more than bad GH (hardness) can. Technically speaking, these aquarium kits are called titration tests.

Titration is a technique widely used in chemistry to measure the concentrations of dissolved compounds. The idea is to have a chemical "reactant" that can capture the ions whose concentration we aim to measure—and a second

[12] Different devices also sometimes use a different ratio to convert electrical conductivity (EC) into TDS. The most common ratios are 0.5 and 0.7, where TDS = ratio × EC, where EC is expressed in units of μS/cm.

chemical compound that will cause a color change when no more free ions are present in the solution. If we know the properties of the reactant and the number of drops required to cause the color of the solution to change, we can calculate how many ions were captured. Combining this measurement with the volume of the water sample lets us, in turn, calculate the concentration of ions in the water. This principle allows us to measure concentrations in HCO_3^-, Ca^{2+}, and Mg^{2+} ions and therefore deduce the GH and KH of water.

In the figure below, I show how a specific property of the solution (the "trigger") changes as more reactant is added, and I also display the range of reactant volumes that actually trigger the color change. The trigger property can be the pH of the solution, but that's not always the case, depending on the specific reactant. In this example, the color shift happens early in the transition, where the measured ions begin to vanish. This means that in this specific example, the best concentration measurement would be obtained by noting the moment at which the color transition has fully completed. Some reactants have this color transition closer to the center of the transition (i.e., the turning point) or toward the end. Unfortunately, most aquarium test kits do not specify which situation we are in, which makes the tests slightly less accurate.

With a bit of research, I learned that the API® GH titration kit uses a reactant based on an EDTA tetrasodium salt. EDTA is a large molecule that can capture calcium and magnesium ions; we want the reactant to slowly deplete these minerals to measure their concentration. The kit includes a buffering compo-

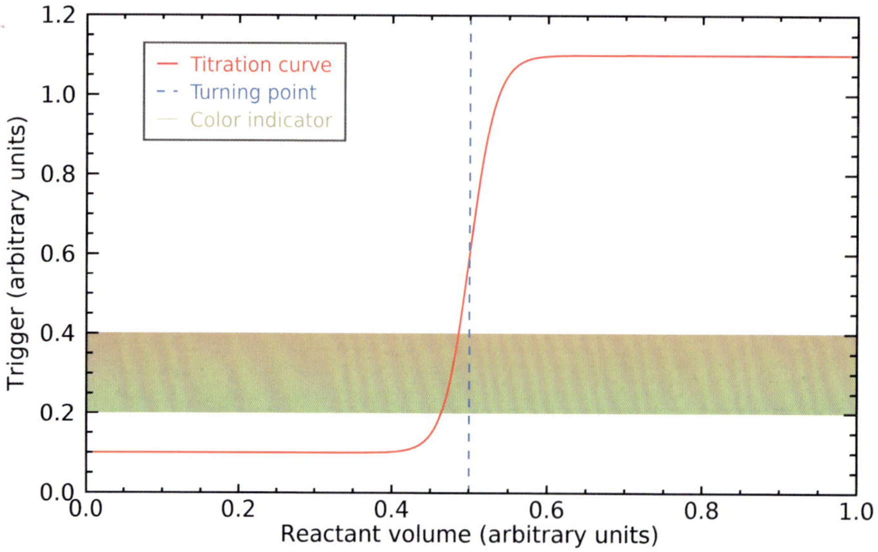

A titration curve (red line) with its turning point (vertical dashed blue line) and the zone where the color indicator shifts (green-to-yellow-shaded region)

nent (triethanolamine) that keeps the water sample's pH very high in order to make the color transition more sudden and improve the measurement precision. The agent responsible for the color shift is not specified, making it impossible for me to determine exactly when the color transition happens with respect to the turning point. Information on the API KH titration kit is even more vague, and I suspect it would be specified only for hazard considerations.

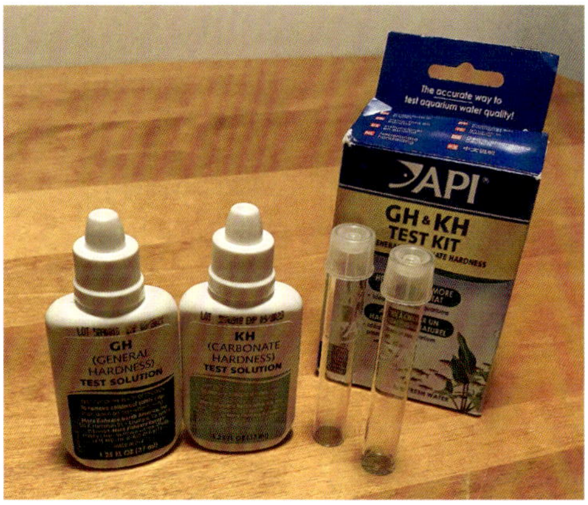

An API aquarium titration kit for GH and KH measurements

When titration tests are performed in a laboratory, reactant volumes can be measured exactly and therefore the tests can allow measurements of concentrations with precisions of about ±1%. We should not expect to achieve this level of precision in a home environment with aquarium kits, however, especially when we rely on counting reactant drops to estimate their volume. Different aquarium kits have different levels of precision, but they are most often limited by the resolution of the measurement: if one drop of reactant corresponds to a hardness or alkalinity of 18 ppm as $CaCO_3$, as is the case for API titration kits, we won't be able to measure concentrations with a better precision if we use the intended methodology. This is quite a large measurement error in the context of coffee brewing, as current recipes have KH values of around 40 ppm as $CaCO_3$.

Usually, these aquarium titration kits require placing 5 mL of water into a small test tube, adding a single drop of reagent, shaking the solution, then looking at the color. Depending on the brand and type of titration kit you are using, different colors will indicate that you have completed your measurement. For the API brand, the GH kits go from orange to dark green and the KH kits go from blue to green and then yellow. Once your solution has reached the color turning point, you can look up a table in the user manual to convert the number of reactant drops to a GH or KH measurement. Typically, every drop counts for 1 German degree,[13] or 17.85 ppm as $CaCO_3$. A slightly more expensive aquarium titration kit made by Red Sea has a resolution of 2.5 ppm as $CaCO_3$ per drop of reagent, but their magnesium hardness test has a much coarser resolution of 82 ppm as $CaCO_3$.

[13] German degrees of carbonate hardness (°KH or dKH) are just different units of total alkalinity often used in aquarium titration kits. One German degree is equivalent to 17.85 ppm as $CaCO_3$.

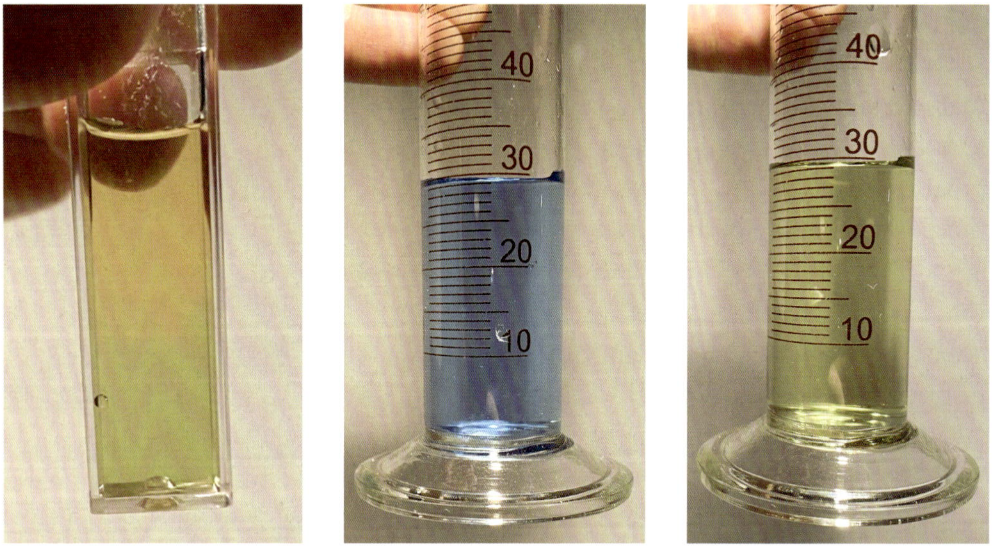

Left: An example of an API GH titration test, in the middle of its orange-to-green transition. Middle and right: An API KH titration test, before (blue liquid) and after (yellow-green liquid) titration is complete.

A few strategies can improve the precision of aquarium titration kits. The first one, which I will call a *stretched titration*, is to simply use a larger volume of water. If you use twice as much water, then every drop of reactant will correspond to half as much measured concentration. This will allow you to easily increase the resolution of the titration—and therefore its precision—but it comes at the cost of using a larger amount of reactant. Furthermore, this may make it seem like using an arbitrary large water volume could allow a barista to measure water properties with an amazing accuracy, but in reality, other sources of error will begin showing up if the sample is stretched too much, and the precision will plateau. For example, the EDTA tetrasodium salt in the reactant will capture magnesium or calcium at different rates, and API kits therefore must make an assumption about the Mg/Ca ratio of your water. This leads to a systematic error in your GH measurement that is smaller than the default resolution and precision of the API kit (approximately 18 ppm as $CaCO_3$), but it can become the dominant source of error if you stretch the water sample enough. For this reason, I recommend considering that your precision is never better than 5 ppm as $CaCO_3$, or 5% of your measurement, whichever value is larger. I have set up a Web calculator[14] that will allow you to determine your concentra-

[14] https://bit.ly/2zMYUTO

tion and precision for stretched titrations with an API GH or KH test. The tool automatically adjusts the precision so that it is not overestimated.

Here are step-by-step instructions for a stretched titration:

1. Thoroughly clean and dry the glass test tube.
2. Wear clean and dry protective gloves.
3. Measure a precise volume of brew water (API's default is 5 mL).
4. Shake the bottle of reactant thoroughly and open it.
5. Add a single drop of reactant to the brew water and stir or shake the mixture well.
6. Look at the color of the solution sideway in front of a white wall in a well-lit environment.
7. Continue adding one drop of reactant at a time to the mixture until you see a color shift happen. Be sure you don't lose count of the reactant drops.
8. Stop counting when the solution has fully completed its color shift, even if this occurs only after you stir or shake the test tube for a dozen seconds.

In cases where water is hard (i.e., GH and/or KH are very high), even a normal titration can require a large amount of reactant, which can quickly become costly. Matters will get even worse if you do a stretched titration on such hard water. For these situations, I recommend a different strategy that is sometimes called a back titration. The idea is to use a small, fixed volume of reactant and then add drops of water to the reactant until the reverse color change is observed. This makes it harder to calculate the concentration in the end, because the aquarium test kits have not done the conversion for you. I would recommend this method for water with either GH or KH above 400 ppm as $CaCO_3$.

The Web tool I have built will allow you to do a back titration and deduce your water properties, but only for the API-brand kits. Before you can use this method, you will need to calibrate your dropper with a milligram-precision scale so that the spreadsheet knows the weight of its average water drop. There is a space just for that in my spreadsheet; simply type in a value, in milligrams, for the weight of 25 drops obtained with your dropper, and the spreadsheet will do the rest.

Here are step-by-step instructions for a back titration:

1. Thoroughly clean and dry the glass test tube.
2. Wear clean and dry protective gloves.
3. Shake the bottle of reactant thoroughly and open it.
4. Put a single drop of reactant into the test tube.

5. Add a single drop of brew water and stir or shake (depending on your test tube) the mixture well.
6. Look at the color of the solution in front of a white wall in a well-lit environment.
7. Continue adding one drop of brew water at a time to the mixture until you see a color shift begin to happen. Be sure you don't lose the count of water drops.
8. Stop counting when you first notice signs that the color of the solution has noticeably begun to shift. The color shift will become more obvious if you add more drops of water, but you should cease counting drop at the moment you first notice the color shift.
9. If you went past the color shift or want a better precision, add another drop of reactant and then keep adding water again. You will just need to update the number of reactant drops in the calculator—and don't forget to always count the total drops of water that you added since the very start.

Back titrations can also be a bit cumbersome because they require counting

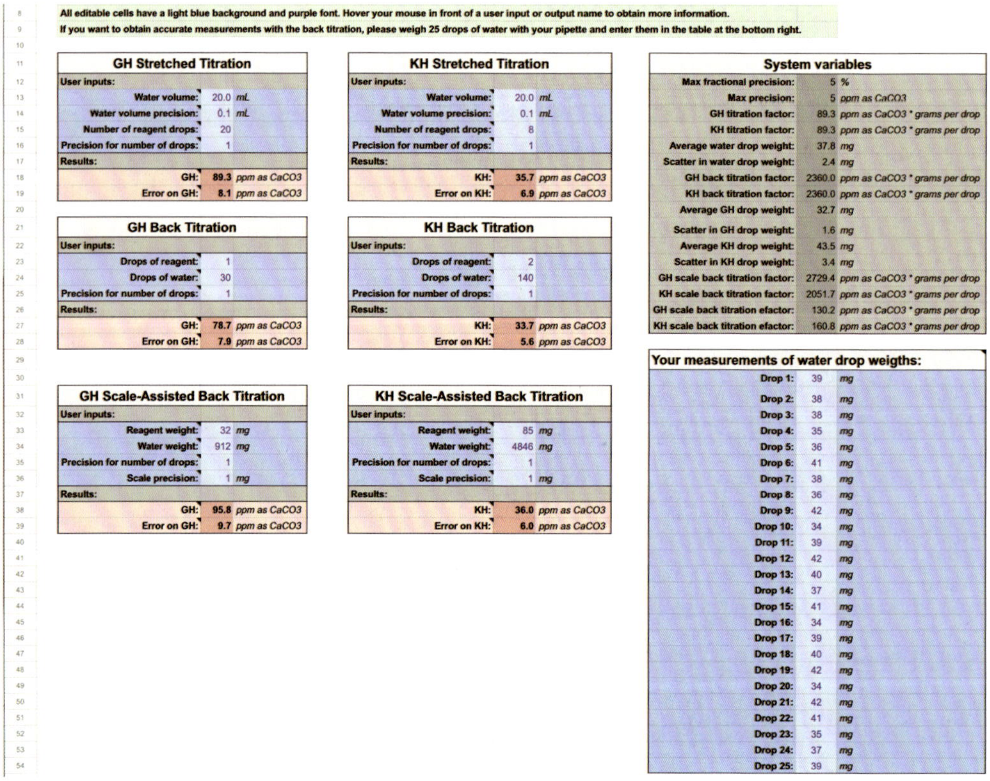

All editable cells have a light blue background and purple font. Hover your mouse in front of a user input or output name to obtain more information.
If you want to obtain accurate measurements with the back titration, please weigh 25 drops of water with your pipette and enter them in the table at the bottom right.

GH Stretched Titration		
User inputs:		
Water volume:	20.0	mL
Water volume precision:	0.1	mL
Number of reagent drops:	20	
Precision for number of drops:	1	
Results:		
GH:	89.3	ppm as CaCO3
Error on GH:	8.1	ppm as CaCO3

KH Stretched Titration		
User inputs:		
Water volume:	20.0	mL
Water volume precision:	0.1	mL
Number of reagent drops:	8	
Precision for number of drops:	1	
Results:		
KH:	35.7	ppm as CaCO3
Error on KH:	6.9	ppm as CaCO3

GH Back Titration		
User inputs:		
Drops of reagent:	1	
Drops of water:	30	
Precision for number of drops:	1	
Results:		
GH:	78.7	ppm as CaCO3
Error on GH:	7.9	ppm as CaCO3

KH Back Titration		
User inputs:		
Drops of reagent:	2	
Drops of water:	140	
Precision for number of drops:	1	
Results:		
KH:	33.7	ppm as CaCO3
Error on KH:	5.6	ppm as CaCO3

GH Scale-Assisted Back Titration		
User inputs:		
Reagent weight:	32	mg
Water weight:	912	mg
Precision for number of drops:	1	
Scale precision:	1	mg
Results:		
GH:	95.8	ppm as CaCO3
Error on GH:	9.7	ppm as CaCO3

KH Scale-Assisted Back Titration		
User inputs:		
Reagent weight:	85	mg
Water weight:	4846	mg
Precision for number of drops:	1	
Scale precision:	1	mg
Results:		
KH:	36.0	ppm as CaCO3
Error on KH:	6.0	ppm as CaCO3

System variables		
Max fractional precision:	5	%
Max precision:	5	ppm as CaCO3
GH titration factor:	89.3	ppm as CaCO3 * grams per drop
KH titration factor:	89.3	ppm as CaCO3 * grams per drop
Average water drop weight:	37.8	mg
Scatter in water drop weight:	2.4	mg
GH back titration factor:	2360.0	ppm as CaCO3 * grams per drop
KH back titration factor:	2360.0	ppm as CaCO3 * grams per drop
Average GH drop weight:	32.7	mg
Scatter in GH drop weight:	1.6	mg
Average KH drop weight:	43.5	mg
Scatter in KH drop weight:	3.4	mg
GH scale back titration factor:	2729.4	ppm as CaCO3 * grams per drop
KH scale back titration factor:	2051.7	ppm as CaCO3 * grams per drop
GH scale back titration efactor:	130.2	ppm as CaCO3 * grams per drop
KH scale back titration efactor:	160.8	ppm as CaCO3 * grams per drop

Your measurements of water drop weigths:		
Drop 1:	39	mg
Drop 2:	38	mg
Drop 3:	38	mg
Drop 4:	35	mg
Drop 5:	36	mg
Drop 6:	41	mg
Drop 7:	38	mg
Drop 8:	36	mg
Drop 9:	42	mg
Drop 10:	34	mg
Drop 11:	39	mg
Drop 12:	42	mg
Drop 13:	40	mg
Drop 14:	37	mg
Drop 15:	41	mg
Drop 16:	34	mg
Drop 17:	39	mg
Drop 18:	40	mg
Drop 19:	42	mg
Drop 20:	34	mg
Drop 21:	42	mg
Drop 22:	41	mg
Drop 23:	35	mg
Drop 24:	37	mg
Drop 25:	39	mg

The Google spreadsheet I created to help users achieve better precision with API aquarium kits many water drops, especially if the water is soft (i.e., it has a low KH and/or GH). My Web tool also lets you use a slightly different strategy, which I call a *scale-assisted back titration,* that allows you to weigh the reactant and water instead of counting the drops. This method can give you slightly more precise results and more realistic measurement errors.

To perform a scale-assisted back titration, first place your titration test tube on a milligram-precision scale and note its total weight. Then add the desired number of reactant drops (remember that more drops will get you more precision) and note the updated weight. Now begin adding drops of water, one at a time. You can pick up the test tube, shake it, and place it in front of a well-lit white wall to make things easier. I recommend noting the total weight at every step (without taring) in case your scale turns off automatically when you take too long. Put the test tube back on the scale when you reached the color turning point, and calculate the total weight of water and reactant that you used. Type these measurements into the Web page spreadsheet to obtain your KH or GH with its measurement errors.

I highly recommend using this method at room temperature, in the range 18–25°C (64–77°F), because otherwise the mass density of the water and reactant drops could change too much, falsifying the mass-to-volume relations that I hardcoded into the spreadsheet. This method is more robust against human error, compared with a regular back titration. The step-by-step instructions are the same as for a normal back titration; you just need to keep track of the total water and reactant weights as you go.

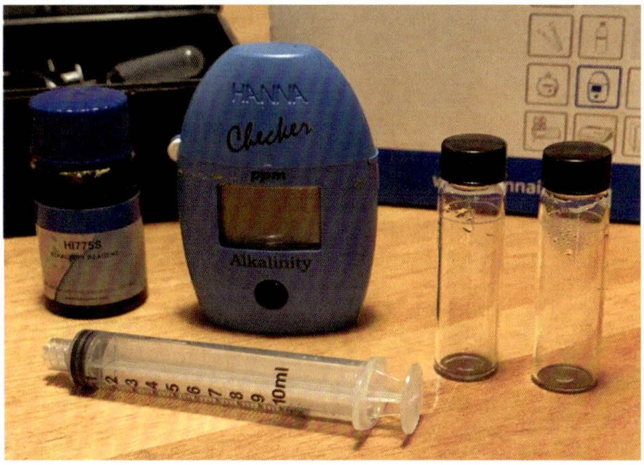

A Hanna Instruments HI775 colorimeter for measuring total alkalinity

Electronic colorimeters provide a more convenient method to accurately measure the properties of water, but they tend to be a lot more expensive, and they also require buying some reactant. One example of such a device is the HI775 colorimeter[15] for KH measurement manufactured by Hanna Instruments.

[15] Be sure to use the model intended for fresh water, not saline water.

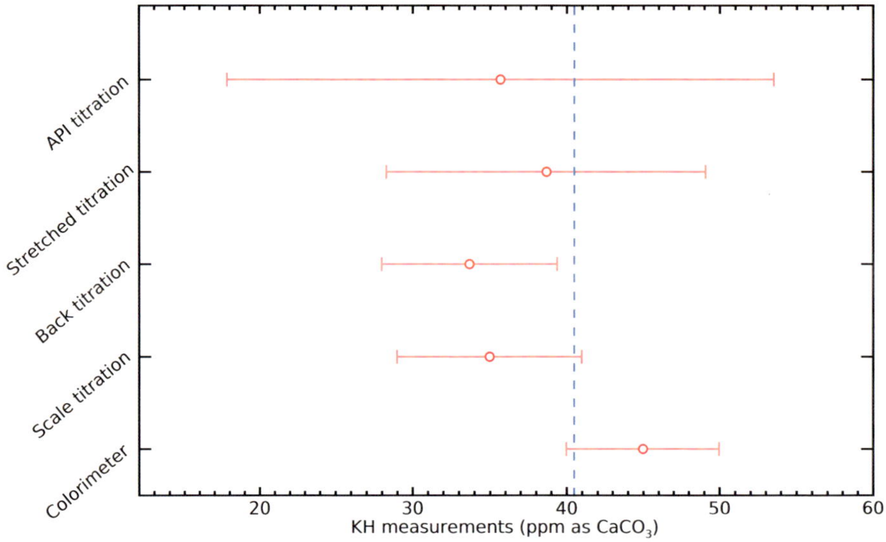

Measurements of total alkalinity made with the different approaches explained here (red circles with error bars), compared with the expected KH of the brew water recipe I crafted (vertical blue dashed line)

If you consider obtaining such a device, it is important to pay attention to their accuracies and to the range of values they can measure. Ideally, you will want an accuracy better than that provided by API titration kits (i.e., smaller than 18 ppm as $CaCO_3$) and a range that extends from about 10 ppm as $CaCO_3$ to a few hundreds of ppm as $CaCO_3$, at least.

Some examples of photometers *not* to get are HI719 and HI720, also by Hanna Instruments, because they are made for measuring calcium and magnesium in extremely low ranges of concentration. Different photometers may also communicate their concentration measurements in different units. For example, the KH photometer HI775 mentioned above provides measurements in units of mg/L HCO_3^-, which you must divide by 1.22 to obtain a measurement in units of ppm as $CaCO_3$. One of the only photometers I found that can measure GH with an appropriate accuracy and range is the HI96735 by Hanna Instruments, but it is much more expensive and more complicated to use.

To illustrate the different methods discussed above, I measured the total alkalinity of a batch of my Rao/Perger brew water with each method. The figure above shows the respective measurements with their error bars, compared with the expected KH value based on the recipe (40.5 ppm as $CaCO_3$).

Most of my measurements were consistent with the expected value within the error bars, but the scale titration and the colorimeter were both accurate *and*

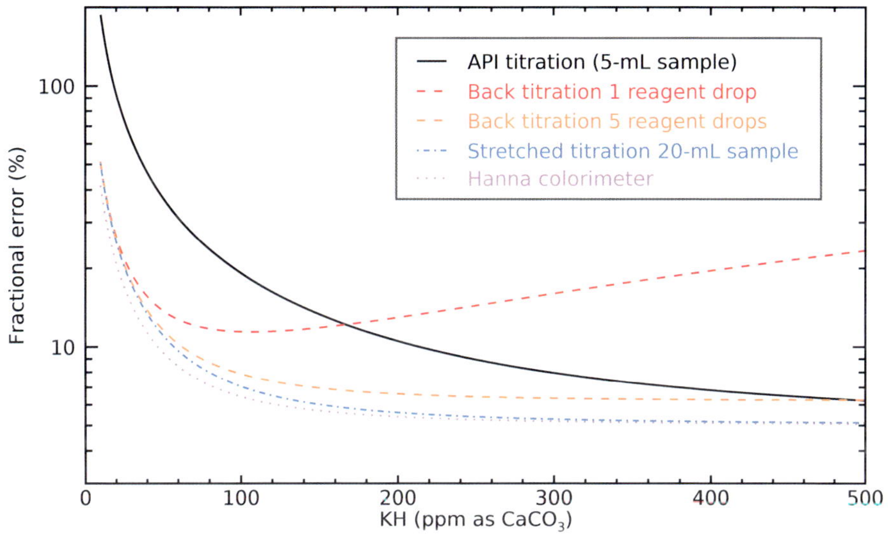

Fractional measurement errors in total alkalinity for different methods discussed

had smaller error bars. The colorimeter is easier to use correctly and takes less time, but it is more expensive.

In order to illustrate the distinct situations where different methods are preferable, I also built a figure (shown above) illustrating the fractional error you would get for KH measurements by each method, depending on the total alkalinity of your water.

The case for GH is similar. The idea is that stretching the API titration kit to a 20-mL sample rather than 5 mL will get you very good precision, at the cost of using more reagent. A back titration will also allow you to reach a similar precision with a fixed 5 drops of reagent for most water compositions, and the colorimeter will reach a very good precision regardless of the water composition. The HI775 colorimeter that I used here also requires a fixed amount of reagent, independent of the water composition.

The figure above shows the *fractional* measurement error—in other words, the error divided by the measured concentration. If you prefer to know the absolute measurement error, please refer to the (similar) figure shown at the top of the following page, although I find it slightly less useful.

To illustrate the cost in reagent associated with each method as a function of water properties, I built yet another figure shown at the bottom of the following page: it clearly demonstrates that back titrations will save you a lot of reagent, compared with normal or stretched titrations, to reach the same fractional error.

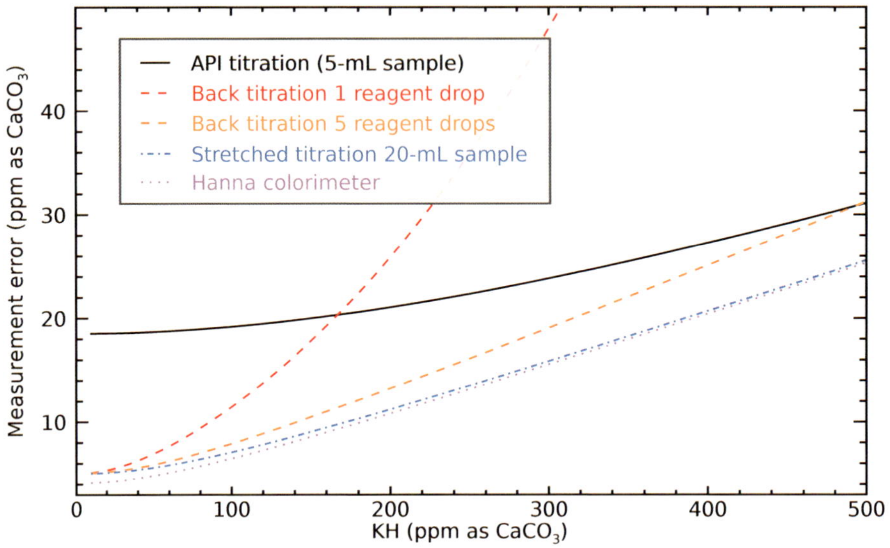

Absolute measurement errors in total alkalinity for different methods discussed

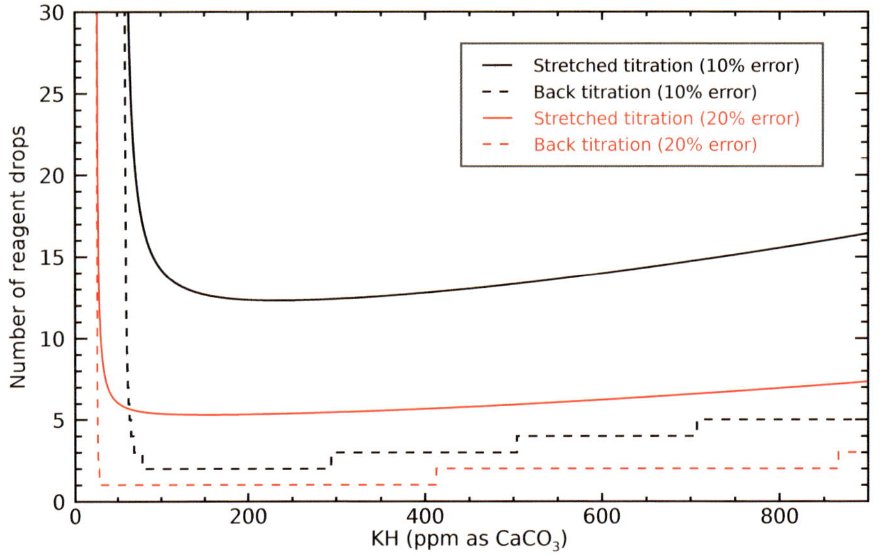

Number of reagent drops required to reach a fractional error of 10% (black) or 20% (red) with the stretched titrations (full lines) versus back titrations (dashed lines). The back titrations require a lot less reactant.

CHAPTER 3

Grinding

It is easy to underappreciate the importance of a good grinder in coffee brewing. One of the most common mistakes I see committed by uninformed home baristas is spending a large amount of money on a home espresso machine while neglecting the quality of the grinder. In this chapter, we will see what effects a grinder can have on the quality of coffee prepared with percolation methods. We will study the concept of a *particle size distribution* and review some of the fundamental physics underlying the fracture of coffee beans. We will use these concepts to understand the important features of a good coffee grinder and how a good grinder will positively impact your coffee brewing.

3.1 Particle Size Distributions

The most common way to represent the performance of a grinder is to display the number of coffee particles it generates per family of particle sizes. Such a figure will inform the reader about the average grind size and its uniformity. Generally speaking, lower-quality grinders tend to generate wider distributions—in other words, they will produce a wide range of different coffee particle sizes all at once—whereas higher-quality grinders will produce narrower or bimodal (i.e., two-peaked) distributions.

The figure on the following page shows an example of a particle size distribution. The full line represents the volume fraction of coffee particles as a function of their size. In this example, those with a size near 300–400 microns form the main peak of the distribution, and they are therefore sometimes qualified as

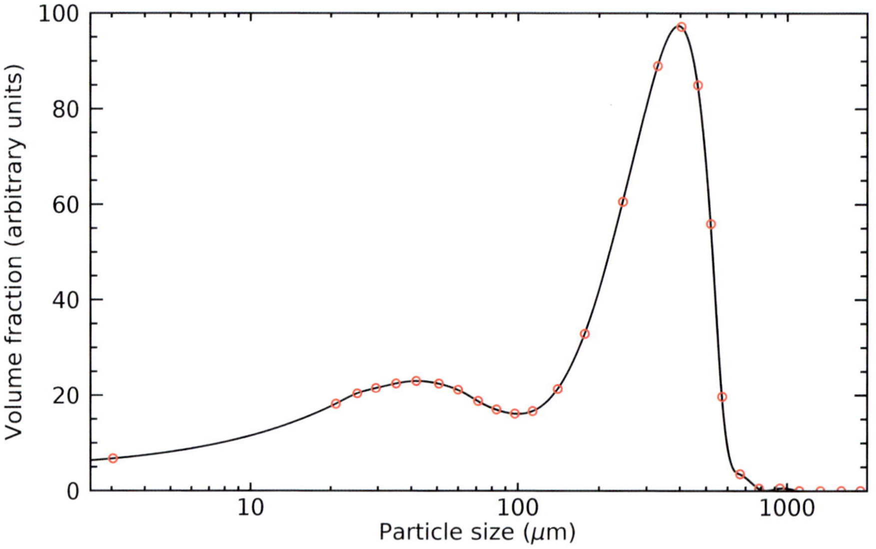

Particle size distribution of a Weber Workshops EG-1 v1 with native burrs *dialed in* for espresso, obtained by John Buckman with a Helos laser diffraction instrument by Mytos and Co.

the target size. The second, smaller bump to the left represents very small particles often called fines. Sometimes, distributions also have some contribution of particles much larger than the target grind size, which is not really the case here; those are often called boulders. As we saw in Chapter 1, smaller particles will extract faster and contribute more of their soluble chemicals, disproportionately affecting the flavor profile.

However, the story does not end there: the particle size distribution will have a significant effect on how water flows through a bed of coffee. As you may have noticed from experience, finer-ground coffee produces a much slower flow. The exact calculations that would allow us to determine this flow rate from the particle size distribution would be extremely challenging because the flow also depends on the shapes of coffee particles, their surface roughness, and exactly how the particles are arranged. For example, tamping the coffee will reduce the space between the particles and reduce the flow of water. For these reasons, scientists typically resort to experimental approximations or computer simulations to calculate the effect of a particle size distribution on the flow of water.

The parameter of a coffee bed that describes how fast water can flow through is called its *hydraulic permeability,* which we will denote as the mathematical variable k. Anderson et al. (2007) define a useful empirical approximation stating that the permeability scales as

$$k = f_{\mathrm{sp}} \left(D_{10}\right)^2$$

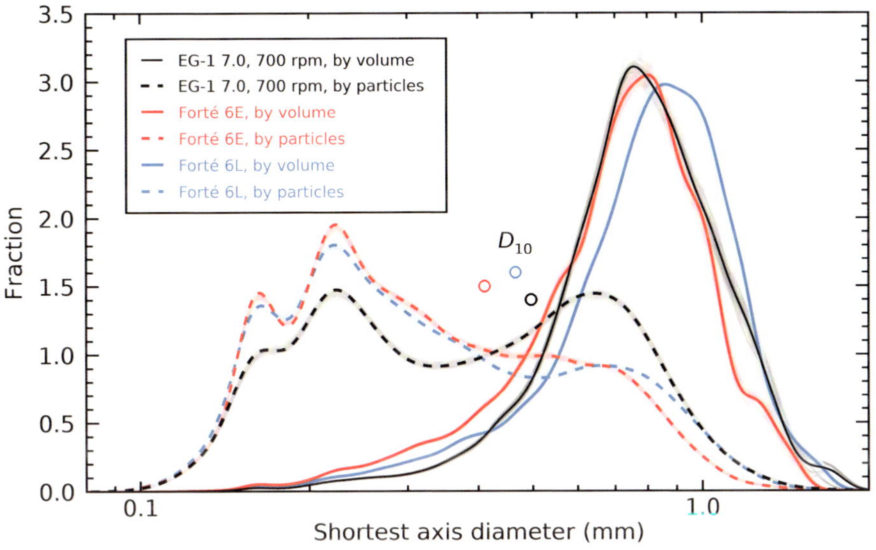

Particle size distributions weighed by particle volumes (full lines) and number of particles (dashed lines). The data were collected from scanned images of coffee grounds analyzed with my Grind Size Application, as described in Chapter 11; the limited spatial resolution of the scanner (400 DPI) prevents detection of fines smaller than about 0.1 mm. Respective D_{10} values are shown as circles. My V60 dial-in for EG-1 v2 was located at 7.0 and 700 rpm, whereas that for the Forté BG grinder was located at grind size 6L, which has a coarser target peak but more fines, resulting in a similar albeit slightly lower D_{10} value indicative of a higher bed resistance and longer brew times. Aligning the target peak by using grind setting 6E on the Forté would generate astringent brews with high bed resistances and enough coffee fines to cause clogging. Pale curves represent random realizations with 80% of the sample size to represent the uncertainty associated with the limited sample size.

where f_{sp} is a quantity that depends on the shapes and packing of particles, and D_{10} is the 10th percentile of particle diameters, as measured by sifting. The concept of a "10th percentile" is relatively simple: if you calculate the cumulative weight of coffee particles in the sieve compartments from the smallest to largest, the sieve mesh size at the point at which you hit 10% of the total weight is D_{10}. This dependency on D_{10} instead of an average size may seem surprising, but it arises from the fact that small particles slow down flow much more than boulders accelerate it.

Generally, a more uniform distribution will result in faster flow if the target size peak remains unchanged, as D_{10} is shifted to a larger value because of the smaller amount of fines. The graph shown above illustrates these considerations. I compared particle size distributions corresponding to my V60 dial-in setup on the Baratza Forté BG grinder, which has 54-mm flat burrs, with particle size distributions of the EG-1 v2 grinder, which has 80-mm SSP ultra-low-fines flat burrs. Because the Forté generates a lot more fines, I needed to grind with a coarser target peak to obtain similar brew times and a pleasant taste profile free of astringency.

In addition to directly affecting the initial bed resistance, migrating fines can also interfere with flow, clogging the pores of the lower coffee bed or the filter. We will explore the concepts of coffee bed permeability and fines migration in more detail in Chapter 4.

As you may have noticed in the previous figures, particle size distributions can be expressed in many different ways, and this can make it a bit harder to interpret them. For example, we can look at the total volume (or weight) contribution of each coffee particle, as was done in the laser diffraction figure at the beginning of this chapter and is very common in the coffee and hydraulic science communities. Because larger particles weigh more, they will be represented more clearly in such a figure, and fines will not be seen as clearly. We can also look at the number of particles of each size, which will tend to make fines much more easily visible because they are typically present in very large numbers. More rarely, figures are drawn to represent the contribution of each particle to the total surface area.

A toy model[16] recently presented by Barista Hustle (Perger 2019, Jan. 3) to fit experimental data showed that the total weight of coffee solubles extracted by a particle of a given size seems to be proportional to the volume of the outermost 200-μm shell of the particle. As we saw in Chapter 1, Moroney et al. (2019) demonstrated that in reality we are not extracting just the outermost shell of a coffee particle, but the slow diffusion of some chemicals inside a particle still makes this model useful in tracking the total amount of solubles contributed by a particle of a given size. This makes it possible to also display a particle size distribution where the contribution of each particle is weighed according to its contribution to the solubles of a coffee cup; this is, in my opinion, a good way to represent how even an extraction will be. The representation by volume is more useful for comparison with literature data. In addition, the representation by number of particles is more useful to clearly outline differences in the production of fines, which can play an important role in how fast water flows through the coffee bed, and in a coffee's potential to clog a filter, for example.

The options do not end here, however. We can choose to display the horizontal axis not only as the particle diameters, but also as the particle surfaces,

---

[16] A toy model is a simplistic model that does not attempt to describe all the details of a phenomenon but still allows us to understand something useful about it.

volumes, or even their diameters across the shortest axis. This last option is probably the best one to use when we analyze imaging data with the intention of comparing it with a particle size distribution obtained from sifting. Think about it: sifting a coffee particle will let it fall through holes as small as its shortest axis, if the particle is shaped like a cylinder. I show examples of each of those possible displays in the horizontal axes on the following pages, using the same data.

By showing you this, I want to give you a sense of how the choice of horizontal axis will affect the perceived shape of the particle size distributions but also make you realize something fundamental about continuous distribution in general: *if you change the variable shown by the horizontal axis, the location of the peak will change.* This observation is more profound than it may sound: if you take, for example, the peak of a particle diameter distribution and calculate the surface that would correspond to a coffee particle of that specific diameter, you will obtain a value that is different from the peak of the same data visualized as a particle surface distribution!

The same thing happens when we display a distribution of how much energy we receive from the sun at each wavelength of light: if you display the distribution as a function of light's wavelength, the peak will be located at a greenish color (~500 nm), but if you display the *same* data as a function of light's

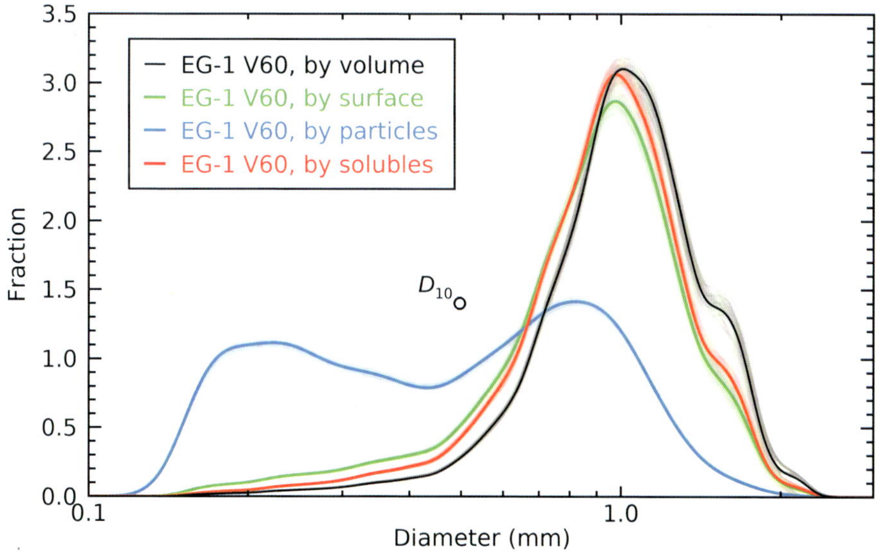

Particle size distributions for the EG-1 v2 grinder with SSP ultra-low-fines burrs at the grind size I use for V60. Each colored curve represents the same data, where each particle's contribution is not weighed (blue), weighed by its volume (black), surface (green), or typical contribution in coffee solubles (red), according to a toy model by Barista Hustle.

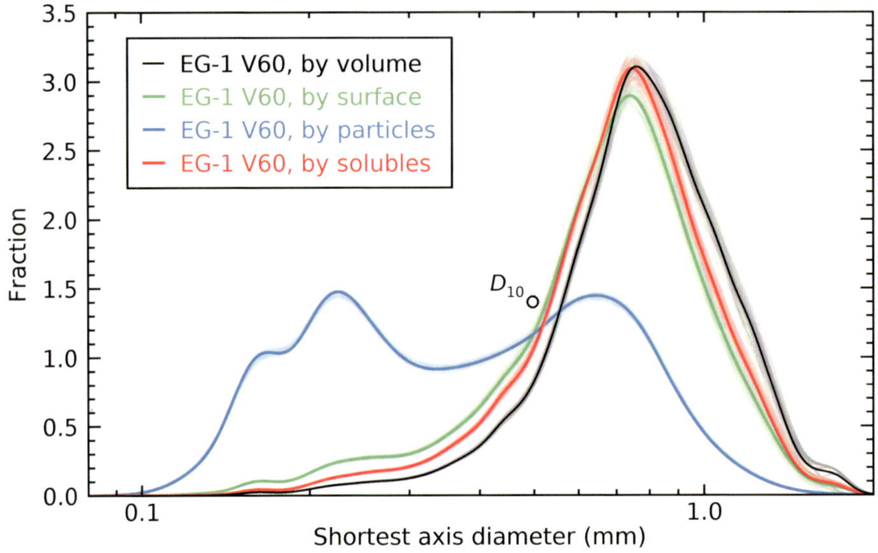

Particle size distributions for the EG-1 v2 grinder with SSP ultra-low-fines burrs at the grind size I use for V60, as functions of the shortest axis diameters. This choice of horizontal axis makes the volume-weighted curve (black) more easily compared with sifting data. This latter curve was also used to determine D_{10}, S_{10} and V_{10} in all figures, to make these quantities as close as possible to permeability versus D_{10} relations based on sifting experiments.

frequency, the peak is located at a frequency that corresponds to an infrared color (~880 nm). This apparent contradiction is related to the fact that the exact peak of a continuous distribution is not a fundamental quantity, and in this analogy, rather than just look at the continuous distribution of light, we actually need to know the response function of the observer's eye to know what color will be perceived. A similar resolution of this apparent problem occurs in coffee: if you choose a particular way to group the data in a particle size distribution—for example, by using a finite set of sieves—you will agree on what sieve set will receive the most amount of coffee regardless of which representation you use to do the calculation.

In addition to all the options listed above, particle size distributions are sometimes displayed with a linear rather than logarithmic horizontal axis. I find this representation less useful because the shape of the distribution of coffee fines dominates this type of representation, and its detailed shape has less direct consequences on coffee brewing compared with the exact location of the target peak or how wide it is. Cumulative distributions are also sometimes displayed instead of direct distributions.

Now that we have seen the many ways in which particle size distributions can be displayed and why they are important, we will have a look the fundamental physics behind grinding—and how grind setting affects these distributions.

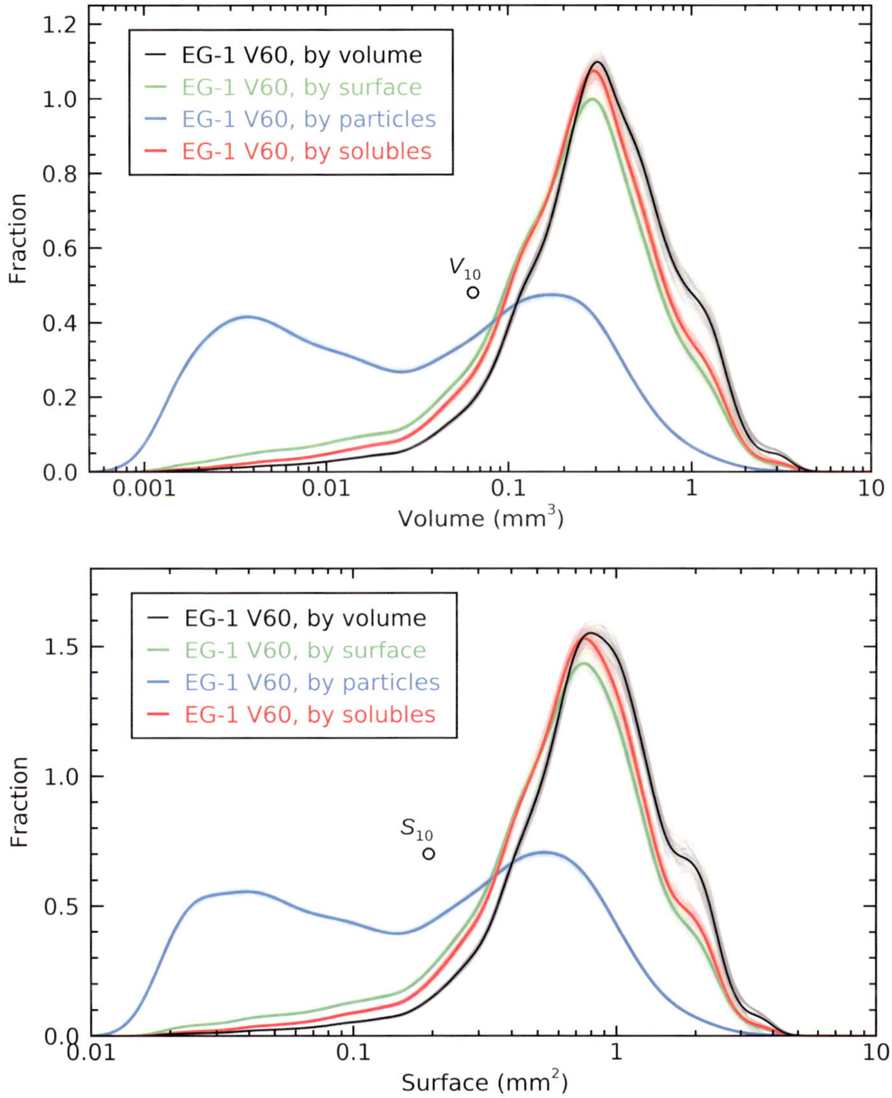

Particle size distributions for the EG-1 v2 grinder with SSP ultra-low-fines burrs at the grind size I use for V60. Top panel: the distributions as functions of particle volumes are displayed. V_{10} is the volume of the 10th percentile particle. Bottom panel: the distributions as functions of particle surfaces, a more straightforward measurement that is independent of particle shapes when analyzing imaging data. S_{10} is the surface of the 10th percentile particle.

3.2 Fracture Mechanics

Two distinct processes can break down a coffee bean into smaller fragments: cutting and crushing. Both of these phenomena aim to deform a material past its *elastic regime* to where it will deform permanently (at what is called the *plastic regime*) and then break apart completely.

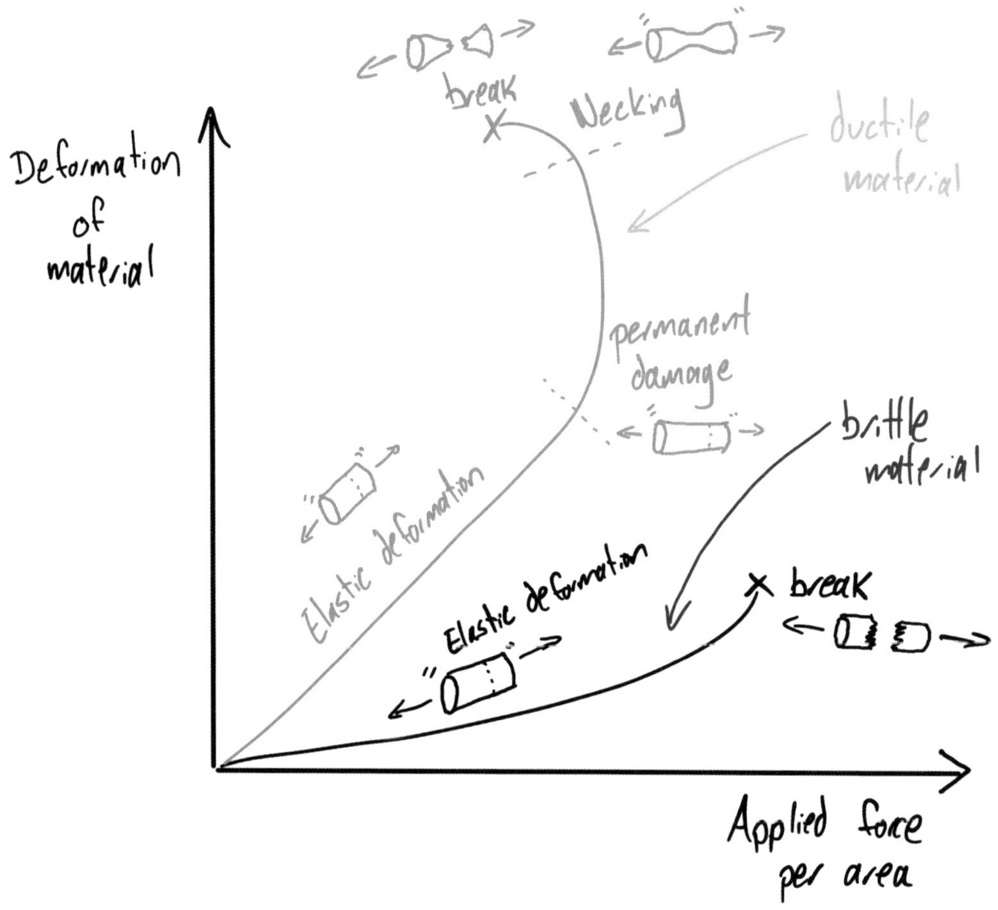

Material deformation as a function of applied force, for brittle versus ductile materials. Ductile materials go through permanent deformation and sometimes a necking phase before completely breaking, whereas brittle materials immediately break after a much more limited deformation phase.

Some materials have a very short elastic regime and will almost immediately break apart when they are cut or compressed; those are qualified as *brittle*. Other, more *ductile*, materials will expand or contract more before reaching their breaking point. This is particularly relevant for coffee beans; darker roasts are more brittle and easier to grind, compared with lighter roasts. Every home barista who has attempted to hand grind a lightly roasted Ethiopian will understand this concept intuitively. If we were to attempt grinding green coffee, things would get even worse, and we would risk damaging most grinder motors because green coffee beans are extremely ductile and require a significant amount of force to reach their breaking point.

In technical terms, cutting results from the application of a *shear force* (i.e., unaligned forces) on an object past its elastic regime, at which point it separates

Cutting a coffee bean with the use of a sharp but misaligned tool

into smaller parts. This is typically done by pressing two misaligned sharp edges onto opposite sides of an object.

Crushing happens differently: an object gets compressed inward until fractures form along lines of weaker material strength. Crushing results in a different outcome because brittle objects tend to fracture into random fractal patterns (Palmer and Sanderson 1991; Zhong, Yue, and Ciancio 2018), as illustrated below.

Such fractal fractures will produce a significant number of fines, compared with the number produced by cutting. As Palmer and Sanderson (1991) demonstrated, the typical particle size distributions that result from fractal fractures follow a *power law,* i.e., the number of particles of a given diameter decreases drastically for larger sizes:

$$\frac{N(a)}{N(b)} = \left(\frac{a}{b}\right)^{-2.5}$$

where a and b are the diameters of coffee particles and $N(a)$ is the number of particles with a diameter a. To put this equation into words, every particle of a given size will be accompanied by about 100,000 as many particles that are 100 times smaller in diameter. This fractal distribution can be illustrated with the particle size distribution on the following page, represented in terms of total particles and total coffee mass (assuming that the coffee beans have a uniform mass density).

In reality, the fractal fracturing of a material will not go on down to infinitely many particles of tiny sizes, as the maths above would predict (Roy et al. 2007). Rather, this pattern will start to break up when the fractures become of the same scale as that of coffee cells, which is about 40 microns in diameter on average (Burke

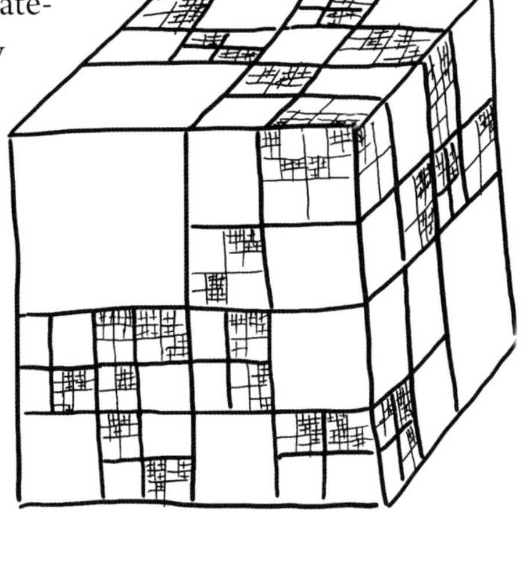

Fractal fracture network. Zooming in on a fracture region will reveal a repeated similar pattern, typical of fractal curves in mathematics.

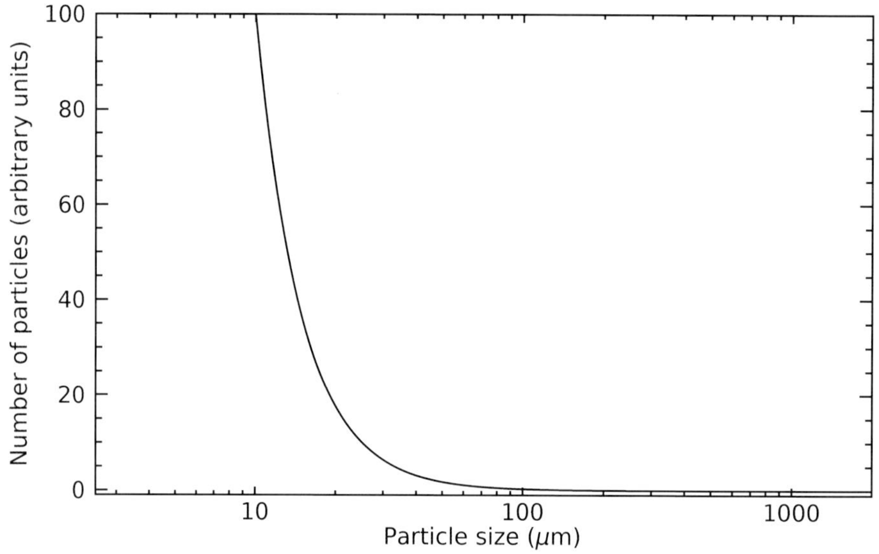

Number of particles for different particle sizes resulting from a fractal fracture network caused by the crushing of a brittle material

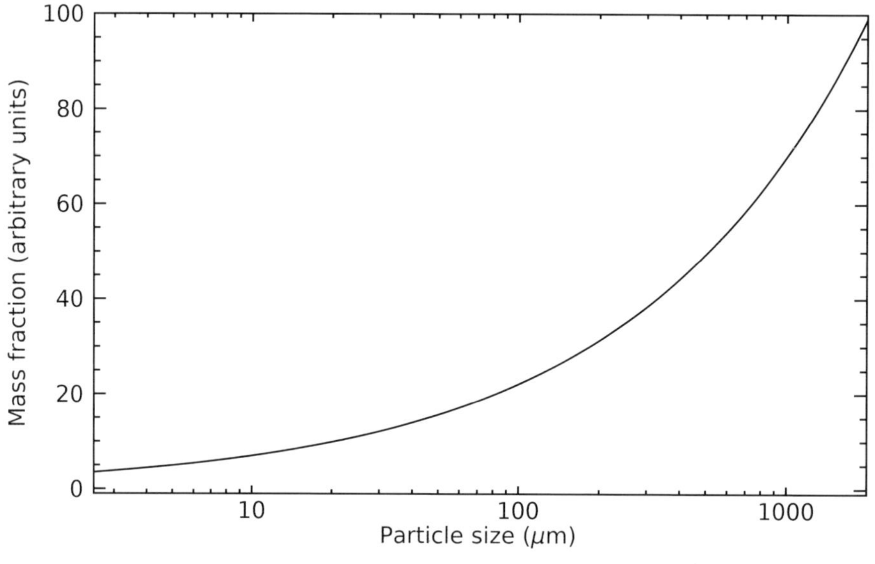

Mass contribution of different particle sizes resulting from a fractal fracture network caused by the crushing of a brittle material

2018). This does not mean that cells will remain intact—far from it. Rather, the fractures will just propagate more slowly and create fewer small particles than what the fractal distribution would have predicted.

Crushing is therefore responsible for the generation of most coffee fines, which, as we will see in Chapter 4, can be detrimental to percolation in various ways. Whether fines are detrimental to taste is a more controversial question and seems to be a matter of preference, but it would be far less controversial to say that we ought to carefully choose the amount of fines we obtain when grinding coffee.

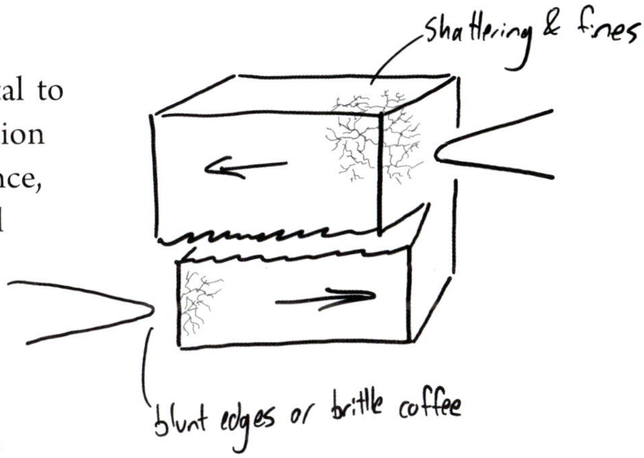

Blunt edges or brittle materials will cause partial crushing during a cutting procedure

Favoring the cutting of coffee beans over crushing is not a straightforward task; it can be achieved by using sharper burr teeth and carefully designing the geometry of burrs. Different coffee beans have different brittleness, as mentioned earlier, and more-brittle beans will always crush more easily and generate more fines, even with the same grinder.

This is where we come to a fundamental limitation of coffee grinding: a grinder will need to perform more cuts to obtain a more uniform particle size distribution, especially if the target grind size is small. However, no cut is ever perfect, and the coffee bean will always get crushed, at least slightly, every time

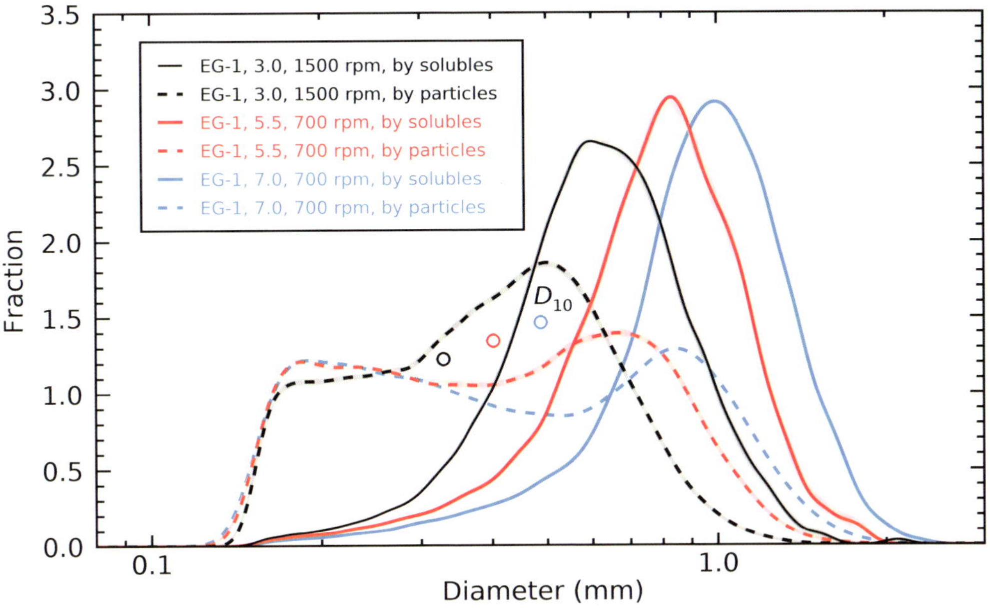

The effect of changing burr spacing on particle size distributions. A higher rpm was required at very small burr separations to avoid causing too much torque on the motor, but it does not significantly affect the shape of the particle size distribution.

the burr teeth attempt to cut it. Therefore, the more cuts a grinder applies, the more fines will be generated. Obtaining a uniform distribution of particles with as few fines as possible requires us to make a compromise between the number of times we choose to cut a coffee bean and the amount of fines we decide to let in.

3.3 Burr Geometry

Two burr types are widely used in the specialty coffee community: conical and flat. Conical burrs are made of two parts, one shaped like the inside and the other like the outside of a cone. Typically, the outer burr is fixed, and the inner burr is rotated with either a crank or motor, causing the coffee beans to be cut and crushed as they fall between the burrs. Gravity will slowly drive the coffee particles downward until they are small enough to exit through the bottom burr spacing. The burr spacing can be changed to obtain a different grind size, typically by lifting up the inner burr.

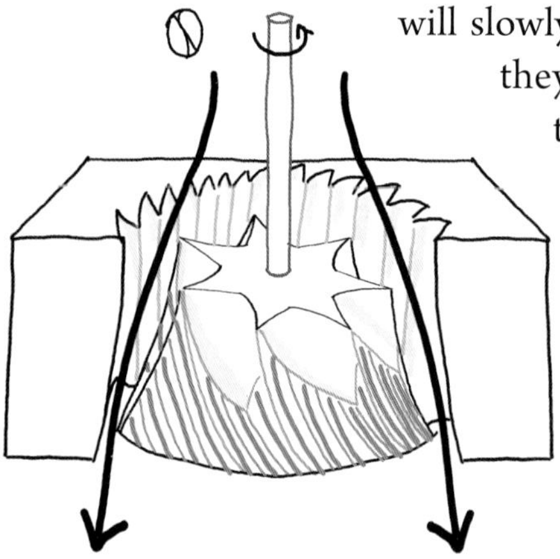

Schematic path of travel for coffee beans ground by a conical burr set

Flat burrs are made of two disk-shaped parts, stacked, with a hole at the center of both disks. Beans are fed through the central hole, and the burrs are spaced according to the desired grind size. After entering through the center, beans travel outward in the radial direction because of the velocity imparted to the beans by the rotating burr. The two burrs can have any orientation because what drives the coffee beans through the burrs is the centrifugal force,[17] rather than gravity. The radius of the disks (minus the central hole) determines the approximate travel distance along which the beans will get crushed or cut.

Each burr type comes with its own set of advantages and inconveniences. Conical burrs are typically cheaper to manufacture, do not require as much force to operate, and can have longer bean travel distances without needing as much engineering, but they have slower coffee throughput (from 1 to 5 grams

[17] The centrifugal force is a "pseudo force" that appears if you imagine a tiny version of yourself placed on one of the rotating burrs. Tiny you will observe beans flying around toward the outer edge of the disk.

per second; von Blittersdorff and Klatt 2017). This makes them particularly suited for hand grinders, which always operate at slow rotation rates. Flat burrs require a much higher rotation rate for the beans to efficiently travel outward, typically between 500 and 1800 rotations per minute (or *rpm*) and can produce much higher throughputs compared with conical burrs (from 1 to 80 grams per second; von Blittersdorff and Klatt 2017). However, large, flat burrs are needed to get a long bean travel path, apply many cuts, and obtain high throughputs, which in turn requires the grinder be operated with stronger motors. The evidence as to which geometry (conical or flat) can produce the most even particle size distributions is still sparse, and as far as I know there are no systematic and quantitative studies available publicly that explore both cases with a range of grinders and teeth geometries to elucidate this question.

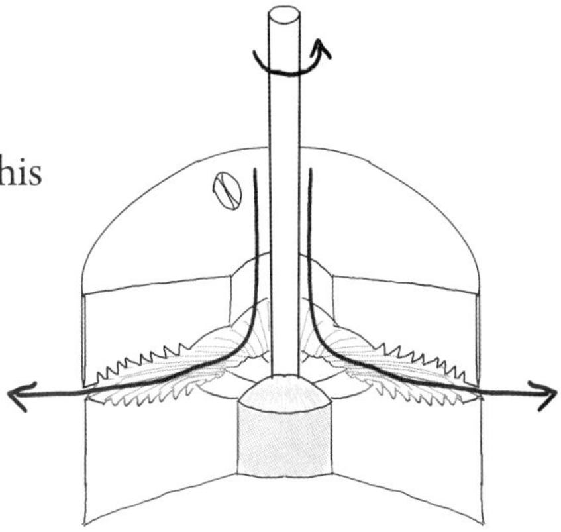

Schematic path of travel for coffee beans ground by a flat burr set

Classical espresso preparation has come to leverage the presence of fines: they are used to reduce the flow of water for an appropriate extraction time, and they can be a factor in what creates the desired mouthfeel of the beverage (Butterworth 2016; Illy and Viani 2004). It is possible to prepare espresso with less fines in the particle size distribution, but doing so will result in a cleaner brew[18] with less mouthfeel. It will also require grinding much finer, using less pressure, or restricting flow with other means to get an appropriate extraction time.

Large grinder burrs typically have three distinct stages along the bean travel direction; each stage features distinctive teeth geometries. First, the *crushing stage* aims at crushing the coffee beans to rapidly decrease the average particle size, which comes with the cost of generating many fines. Second, the *cutting stage* aims at reducing the average size of coffee particles to the desired grind size and make the particle size distribution more uniform without generating as many fines. Third, the *finishing stage* can use reduced spacing between the burr teeth to force any remaining boulders to be cut or crushed. In the most extreme cases, the finishing stage might be very flat, which will prevent any boulder from exiting the burrs. I see this stage as a "last resort" to remove most boulders, at the cost of generating more fines from their crushing.

Blade grinders are also sometimes used to grind coffee, but I do not recommend them because they make it much harder to obtain a repeatable grind

[18] with less body and more easily distinguishable origin characteristics

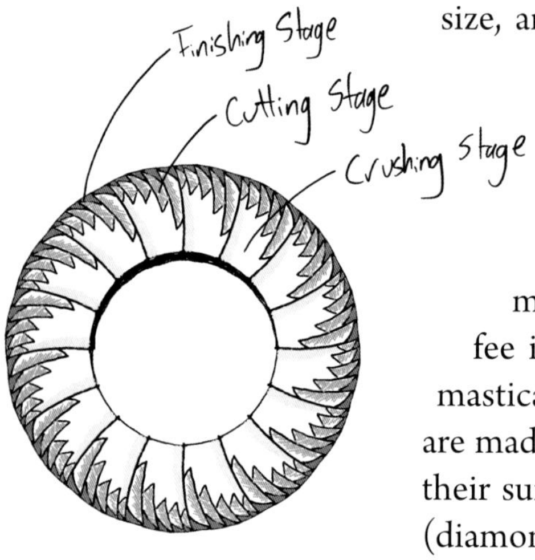

Illustration of the three grinding stages of a flat burr

size, and they generate a large amount of fines and boulders. The average grind size that they produce depends on how long the grinder is operated and how much it is shaken during that time.

Ghost burrs are another, less common, type of grinder burr used in the coffee industry. They are sometimes also called masticating burrs, peg disks, or false burrs. They are made of flat disks with several square pegs on their surface, which can be either blunt or sharp (diamond shaped). These burrs cannot grind as fine as other conventional flat or conical burrs, and they are therefore mostly used specifically for very coarse grinding. They generate more fines because they favor crushing the coffee particles, especially those with blunter teeth (Rhinehart 2016; Sunergos Coffee 2012), but there is anecdotal evidence that they generate a uniform target size peak.

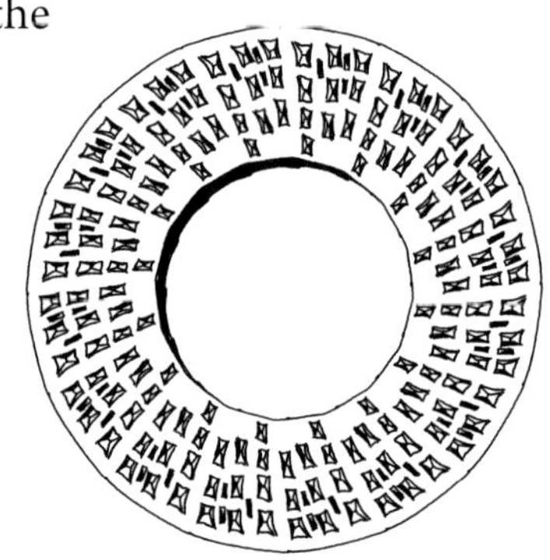

Illustration of a ghost burr, which uses square pegs to crush or cut coffee beans

3.4 Burr Alignment

Poor burr alignment is a common problem with both commercial and home flat burr grinders. If the burrs are not perfectly concentric and coplanar, the teeth will not perform as intended, and the particles coming out from different directions along the circumference of the burrs may have different grind sizes. This will result in a wider particle size distribution and can even prevent very-fine grinding.

One method I like to use to quickly assess burr alignment in an approxi-

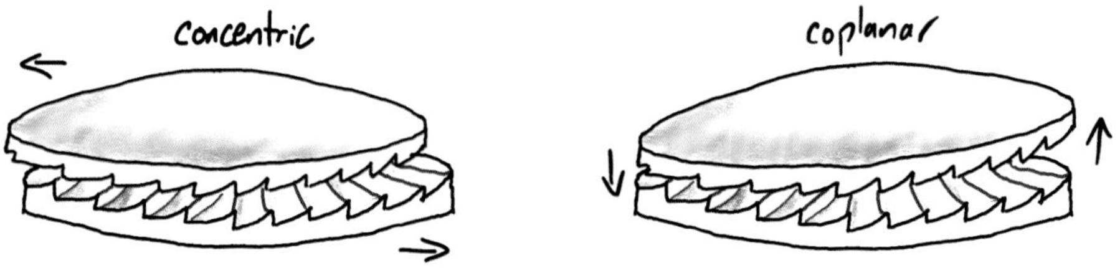

Types of burr misalignment

mate manner is to grind as fine as I can and then verify whether the resulting coffee powder can hold my fingerprint.

There are various methods available online to realign grinder burrs, but their efficiency depends on grinder models, and I will not enter this particular deep topic of discussion here. Some private companies offer burr alignment services for the less manually inclined baristas.

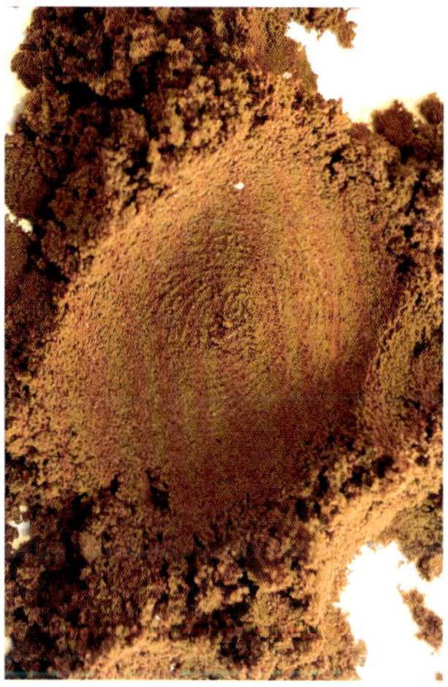

Coffee ground fine enough to retain a fingerprint

3.5 Rotation Rate

Most coffee grinders operate at a fixed, predetermined motor speed, but some high-end grinders offer the possibility to choose the rotation rate. Rotation rate will affect the impact velocity of cuts, and the frequency of impacts, but it will also change the travel velocity of beans through the burrs. In my limited experience, a faster rotation results in a finer grind size when the burr spacing is left unchanged, but this may depend on burr design. Grinding at lower rotation speeds will place more strain on the grinder's motor because the momentum of the burrs will not contribute as much to maintaining the rotation rate, and the efficiency of the grinder in terms of the total amount of coffee that can be ground per minute will also be reduced. This will, in turn, cause the coffee to remain between the burrs longer and heat up more as a result, which can negatively affect taste (Klatt 2015).

In the figure at the top of the following page, I show the effect of grinding rpm that I observed with the Weber Workshops EG-1 v2 grinder with SSP ultra-low-fines burrs. We can see that the faster rpm resulted in a slightly finer grind size. We cannot see particles smaller than 0.1 mm in these distributions because they were built with resolution-limited scans and the grind size software application (discussed in Chapter 11). I was not able to distinguish between grind size 4.1 at 700 rpm and grind size 4.7 at 1500 rpm either by taste or by extraction yield measurements, suggesting that this shift of 0.6 in grind size compensated for the different rpm and the effect on fines was not dramatic—at least, not with this combination of grinder and burr geometry.

Because rotation rate affects the average grind size, any inconsistency in the rotation rate during grinding can decrease the uniformity of the resulting particle size distribution. For example, starting a grinder while some beans are already against or between the burrs will result in a wider and coarser particle

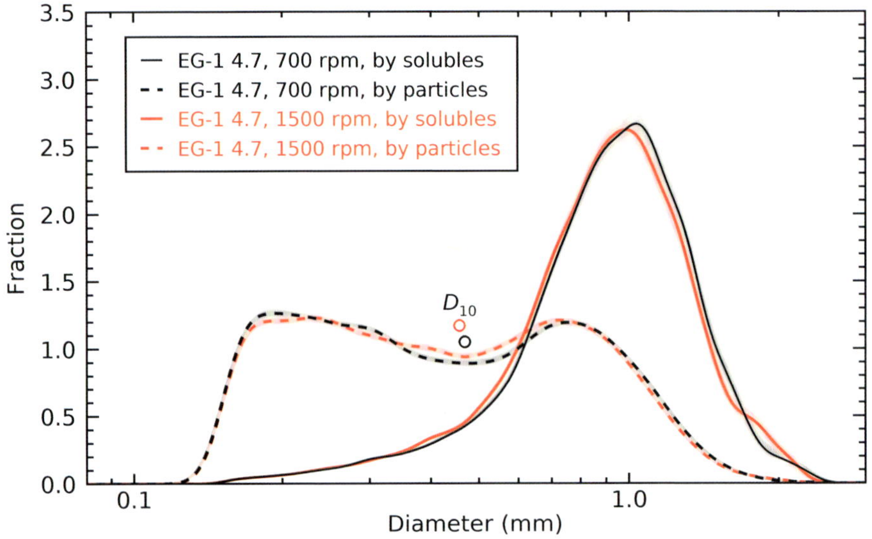

Particle size distributions resulting from grinding at different rotation rates with the Weber Workshops EG-1 v2 and SSP ultra-low-fines burrs. Faster rotation rates resulted in finer average particle sizes and a similar grind quality.

size distribution because the first beans will be ground at a lower rotation rate as the burrs gradually speed up to their final rotation rate. It is therefore desirable, when possible, to turn on the grinder while it is empty and then pour the coffee beans in after the rotation rate has stabilized.

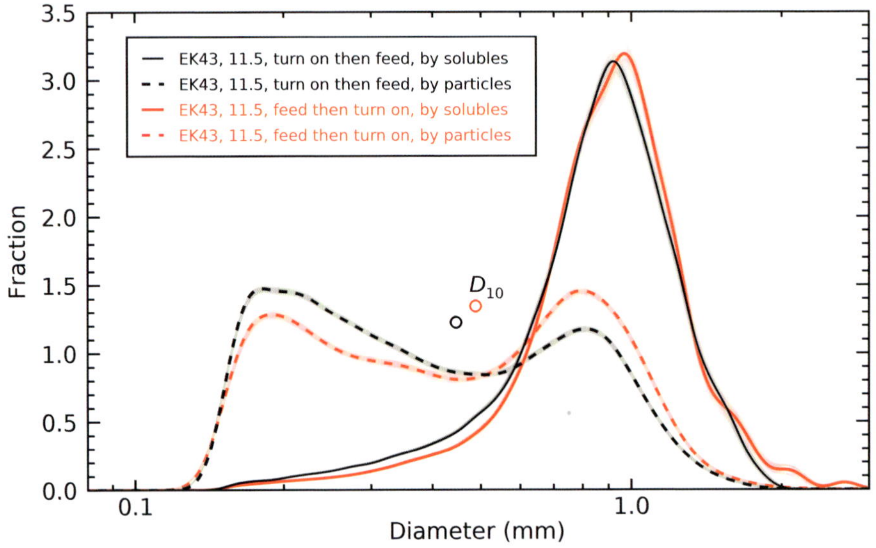

Particle size distributions resulting from coffee ground with a Mahlkönig EK43 with original burrs at variable rpm because the motor was started with beans already between the burrs (red), versus a dose ground with the motor stabilized before the beans were inserted (black). The two cases resulted in different average particle sizes.

3.6 Wear and Seasoning

As the teeth of a grinder slowly wear, they will get blunter and will, over time, begin to favor crushing instead of cutting. As a consequence, the grinder will gradually generate more fines and a less-even grind size. The spacing between the burr teeth will also become slightly wider at a fixed grind setting, and the barista will therefore gradually have to adjust the grinder to a finer setting to obtain the same average grind size. This will typically be noticeable when the extraction yield of the brewed coffee gradually goes down at optimum dial-in (Rao 2015), providing an indication that burrs need to be resharpened or changed. Burrs typically begin to wear out after having ground about 4,000 to 10,000 kg of coffee, but different grind sizes, rotation rates, and coffee origins will wear down burrs at different rates. The burrs' material and coating will also affect their lifetimes, and the exact time appropriate for resharpening will therefore be specific to the characteristics of the grinder and how it is used. I suspect that the presence of small stones often undetected in coffee bags may contribute to the sudden wearing of grinder burrs. Since I began inspecting my dose of coffee to identify defective beans before feeding them to the grinder, about 6 months ago, I have already caught two stones in two popular specialty roaster bags, which would have otherwise made it into my grinder.

At the start of a burr set's lifetime, a reverse effect is observed, as its efficiency and grind quality will gradually get *better* as the first few kilograms of coffee are ground. This is likely due to the wearing out and smoothing of micro-defects in the initial burr shapes; this is generally called burr seasoning, or break-in. The exact amount of coffee required to fully season coffee burrs is usually not quoted by manufacturers and will depend on many variables, much like the rate of wear. In my experience, about 12 pounds (5.4 kg) of beans were required to season a Baratza Forté BG with 54-mm steel burrs, and 24 pounds (10.9 kg) were needed to season a Weber Workshops EG-1 v2 with 80-mm SSP ultra-low-fines burrs. I seasoned the latter using test roasts of mixed origins with grind setting 5.0 at 800 rpm[19] and sampled the particle size distribution every 2 pounds at grind size 8.5 at 700 rpm with a light-roasted washed Burundi Bourbon Buziraguhindwa Remera Shade #2 Lot #BI-2018-901[20] (crop year 2018). These samples of ground coffee taken at every 2 pounds during the seasoning step allowed me to reconstruct how the particle size distribution changed during seasoning.

---

[19] The first two pounds were seasoned at grind size 8.5 and 700 rpm, and a faster 1000 rpm rotation was used past the 12-pound mark. Be mindful that quickly grinding a large amount of coffee indoors can generate dangerous amounts of carbon monoxide.

[20] https://hub.cropster.com/store/listings/5139

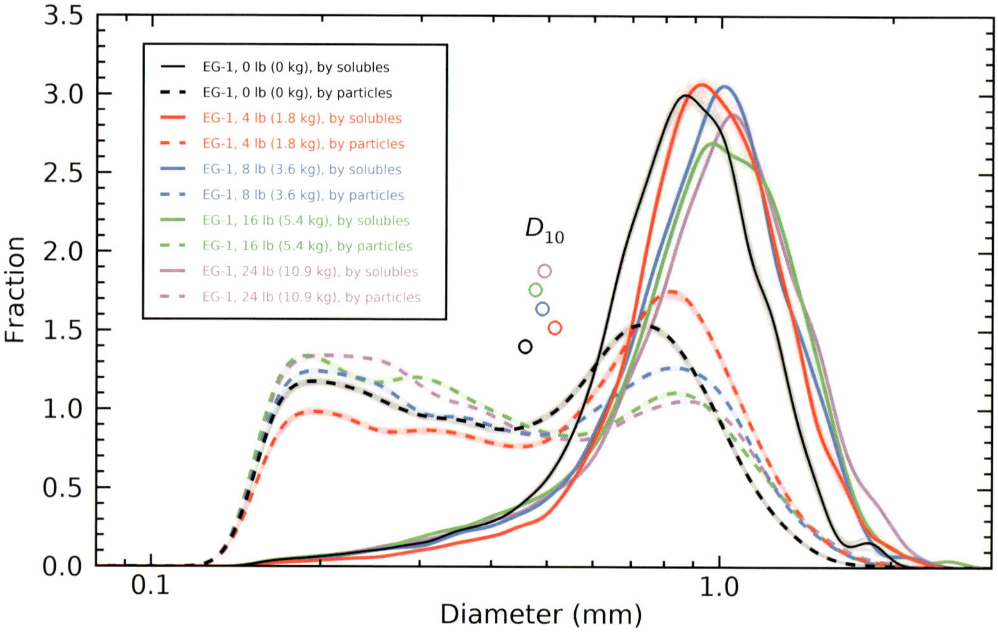

Particle size distributions of the Weber Workshops EG-1 v2 grinder with SSP ultra-low-fines burrs at grind size 8.5 and 700 rpm during seasoning. The particle size distribution moved to coarser sizes on average for the same setting as the seasoning progressed; this is likely caused by an initial wearing of the burr teeth.

As the teeth wore down, the grind size for a fixed burr spacing became gradually coarser. As a consequence, baristas should expect to have to change their optimum grind size often after installing new burrs or a new grinder. It is also useful to remember that the overall grind quality will slowly improve during this stage, and it will later degrade as the burr teeth become blunter (e.g., Socratic 2015, Dec. 3).

3.7 Bean Properties

As we mentioned earlier, the brittleness of coffee beans will have a strong impact on the amount of fines generated by any grinder. The brittleness can likely be affected by a myriad of variables, including the beans' moisture content, roast development, growing altitude, processing, and *variety*. Beans with more moisture tend to be more ductile and therefore generate a smaller amount of fines (Klatt 2015). Dark-roasting of coffee beans weakens their structural integrity and makes them drier and more brittle. Decaffeination processes tend to cause a similar effect, although much more drastic; decaffeinated coffee typically needs to be ground coarser as a result, to avoid clogging paper filters.

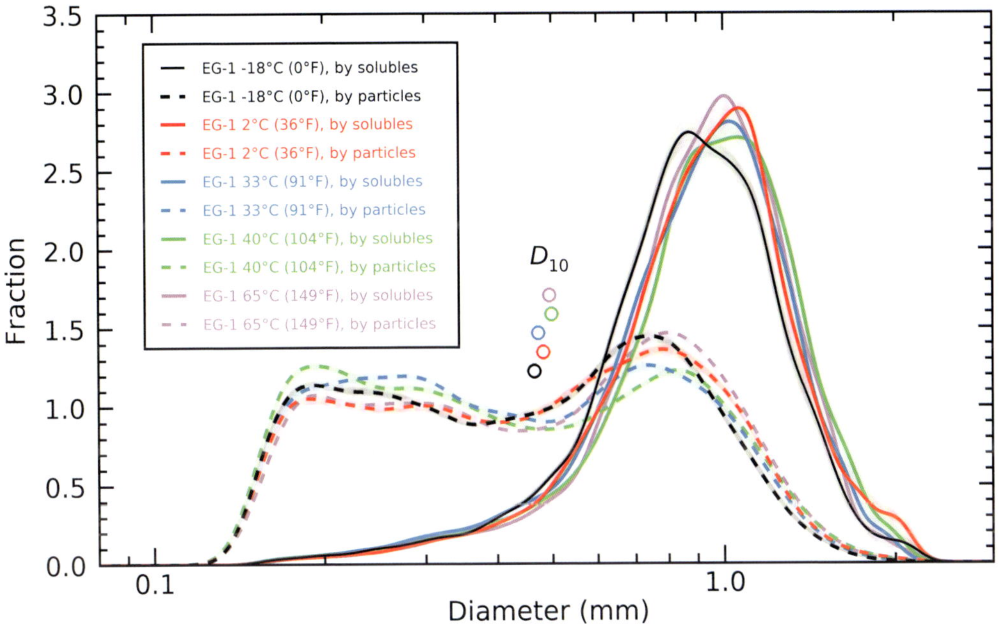

Particle size distributions for the same coffee ground at different temperatures, using a freezer or a microwave and an infrared laser thermometer to measure the temperature immediately before grinding. I observed no clear correlation between temperature and grind size, except for the frozen coffee, which ground slightly finer. It is likely that differences exist below 100 μm.

Another property that affects how coffee beans are ground is their temperature. Colder objects generally tend to be more brittle. Consequently, coffee shops often need to grind gradually finer for espresso to maintain shot time as the shop becomes busier and the ambient temperature rises, making the beans more ductile (e.g., Cameron 2016). The grinding of frozen beans will have a reverse effect and will tend to produce more fines. This led Uman et al. (2016) to posit that grinding cold beans is desirable for espresso, because favoring the generation of fines is generally regarded as positive in espresso preparation (e.g., see Illy and Viani 2004). It is likely that this is *not* desirable for filter coffee, because fines will interfere with percolation (see Chapter 4), promote clogging of filters (see Chapter 5), and contribute flavor compounds that are usually avoided in filter coffee (Rao 2017).

Limited experiments have been done in which coffee beans are microwaved just before grinding in an attempt to warm them and reduce the production of fines for filter preparation (Hoffmann 2019, Aug. 15). In my personal experience, this had a negligible effect on extraction yield, and it negatively impacted taste; this is similar to the findings of Klatt (2015). Preliminary work showed that the taste profile may be impacted by the loss of volatile aromas during either the heating or the grinding with warm beans, or possibly to chemical reactions, although this is still an open question (Baggenstoss et al. 2010; Klatt 2017).

This could mean that frozen coffee may retain more volatile compounds in the brew and, in principle, offset the downside of fines production, even for the preparation of filter coffee, but confirmation of this would require a careful sensory analysis experiment. From my limited experience, I have enjoyed filter coffee ground frozen at least as much as coffee ground at room temperature and I have not observed a clear drop in average extraction yield.

It is important to note that typical changes in room temperature (e.g., 20°C to 35°C, or 68°F to 95°F) will not cause significant expansion or contraction of steel burrs. Steel has a thermal expansion coefficient of 1.2×10^{-5} (Callister and Rethwisch 2007, Oberg et al. 2012), meaning that a typical 9-mm-thick flat burr will expand by only 1.7 μm for a 15°C (27°F) change in room temperature. Combining the effect of both burrs makes up for less than the smallest grind size units on most grinders (typically, 5–25 μm). This had initially been posited as a possible explanation for the need to change grind size when the ambient temperature changes, although it would produce an effect *opposite* of what is observed: expanding burrs at warmer temperatures would *decrease* the spacing between burrs and result in a finer grind size and longer brew times (von Blittersdorff and Klatt 2017). Note that depending on the exact grinder design, an expanding motor shaft may affect the burr spacing by a significant amount.

3.8 Heat Management

One goal of a coffee grinder is to achieve good grinding efficiency and quality—without increasing the temperature of either the beans or burrs. This avoids the generation of static electricity, extraction of oils from the beans, and negative impacts on the taste of coffee (Klatt 2015; Uman et al. 2016). Achieving this goal requires the use of materials that are efficient at conducting and quickly dispersing heat away from the grinding surface, as well as avoiding a situation where motor heat is gradually transferred to the burrs, which tends to happen during heavy use. Placing the motor far from the burrs is a strategy that can reduce this effect. Specific designs in teeth geometry can also help mitigate the accumulation of heat: specifically, thin edges and sharp teeth made of either technical ceramics, alloys, or steels are good for heat dispersal (Loeza 2015). Grinding finer will, unsurprisingly, generate more heat, and grinding at higher rotation rates will decrease the production of heat because beans will travel faster through the burrs and have less time to accumulate heat (Perger 2020; Klatt 2016).

3.9 Static Electricity and Clumping

Another common problem coffee grinders face is the generation of static electricity, which can lead to the clumping of fine coffee grounds. Clumping can reduce the evenness of flow and extraction during coffee brewing, much like the effect of a more uneven particle size distribution. This tends to be more of a problem when grinding very fine, as required for the preparation of espresso, but it can also happen at coarser grind settings, depending on the situation. Coffee oils can also be a factor in the generation of clumps: darker roasts and higher grinding temperatures can therefore be expected to make clumping worse.

The mitigation of static electricity is an infamously hard problem to tackle. The material properties of the grinder, burrs, burr coating, and the container in which the coffee is ground can have a significant impact on the amount of static electricity that the grounds will accumulate. The amount of friction between coffee beans and burrs is also impacted by grind size, bean hardness, and rotation rate, also affecting the production of static electricity. Even variables such as countertop material and the clothing of a barista have been noted to affect the production of static electricity (Anderson 2014). The presence of moisture is generally good at preventing static electricity, hence grinding coffee beans with a higher moisture content or in a more humid environment can help prevent the accumulation of static. Because of this, a common practice known as the *Ross Droplet Technique* is to add a few droplets of water to the coffee beans just before grinding, although this is more common in espresso preparation. Some have noted that waiting for a few minutes after grinding can also allow the static electricity to dissipate (Anderson 2014), and some grinder manufacturers recommend shaking the ground coffee in a tumbler to allow static electricity to dissipate (e.g., Weber 2020).

Static electricity can cause clumps to form in finely ground coffee.

3.10 Sifting

One available solution for narrowing a particle size distribution is the sifting of coffee grounds. This is achieved by shaking the ground coffee over one or more perforated layers that have uniform holes of the desired size. Once the sifting is complete, coffee particles finer than the desired range are usually discarded, and the larger particles are either reground or discarded.

While sifting can be efficient in narrowing a particle size distribution, it takes several minutes and requires regular cleaning of the sieves, making it highly impractical for high volume situations and quite bothersome, even at home. Sifting coffee grounds is generally hard to replicate: thorough shaking is required to efficiently move all small particles through the sieve holes, especially those particles of a size close to that of the sieve holes. This is made worse by elongated particles that can require a precise orientation for fitting through a hole and by the tendency of fines to stick to larger coffee particles.

These effects make the process of sifting imperfect and inconsistent. The resulting particle size distribution will reflect how long the coffee grounds were sifted and how hard the sieve was shaken. Depending on the evenness of the initial particle size distribution, sifting may also waste a significant fraction of coffee grounds, making this option not only less efficient but more costly and less environmentally friendly.

Coffee fines sticking to larger particles, preventing efficient sifting

Sifting is further complicated by the crucial need for dialing-in a coffee recipe to the specific sifting methodology. To my surprise and dismay, some coffee professionals suggest sifting away either fines or boulders without adapting anything else about a brew method. Any change a barista brings to the particle size distribution may require dialing-in their grinder from scratch to determine the appropriate grind size. If boulders are sifted away, the optimum grind size will likely become slightly coarser; removing the fines will have a more drastic impact on flow and will typically require grinding much finer. One possible exception to this general rule is the partial sifting of fines from a specific coffee bean that generates an excessive amount of fines (typical of some Ethiopian coffees, for example).

3.11 Single Dosing

Most commercial grinders come with a receptacle in any of various sizes, termed a *hopper*, that holds coffee beans and feeds them to the grinding burrs. The weight of coffee beans in a hopper forces the bottom beans through the burrs and prevents them from bouncing back up, an effect often called *popcorning*. How much push is imparted by the upper beans can be partly controlled by changing the angle at which a grinder feeds the burrs: a more inclined feeding angle will reduce this pushing effect. These considerations can have an impact on the resulting particle size distribution. To understand this better, it is useful to remember that a coffee bean about to enter the spacing between the grinder burrs will encounter a spacing size that is dependent on the alignment of the teeth as the burrs rotate, even if the burrs are aligned.

If the beans are not forced through, they might bounce around until they encounter an adequately large spacing. If forced through, they will instead get partly crushed and then go through the remainder of the grinding path, having withstood more significant crushing during their initial entry. Placing only a single dose of coffee in an otherwise empty hopper can cause both scenarios to happen; the first part of the dose will initially be pushed through, whereas the last beans will "popcorn" (i.e., bounce around before going through the burrs) and be ground to a slightly different size.

These considerations can be further complicated when coffee particles interact with each other during their travel through the burr spacing: random collisions between densely packed coffee particles may cause them to remain between the burrs longer or eject faster, causing the resulting grind size distribution to be wider. I am not aware of any publicly available data that show how important this effect is. Other variables can also affect the degree to which coffee beans popcorn in a grinder: finer grind settings that result in smaller burr spacing, faster rotation rates, and burr geometries with narrower crushing stages can all increase it.

3.12 Retention

Most coffee grinders have cracks, screw holes, or spacings that can store ground coffee during normal use. The coffee grounds will accumulate until the grinder is

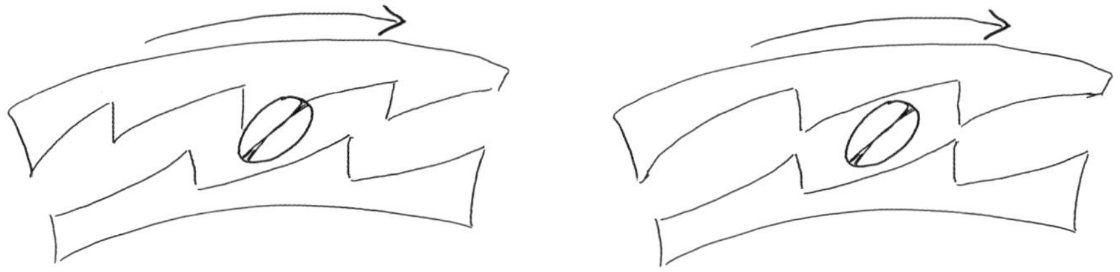

View of a rotating flat burr set seen from the central hole as a bean enters the burr gap. A coffee bean may bounce back if the burr spacing is small and the teeth are not aligned.

dismantled for a deep cleaning, which can cause most grinders to release a small amount of older coffee during grinding as the retained grounds get replaced by fresher coffee. This effect, called grinder retention, is undesirable, especially in situations where coffee beans are changed often or when small volumes of coffee are ground every day, allowing the retained coffee to stale. In a high-throughput environment, purging the grinder by grinding a small amount of fresh coffee and throwing it away helps fill up retaining spaces with fresher coffee. This will reduce cross-contamination from older retained grounds. An increased inconsistency in the dose produced by a grinder can be an indication that it needs to be cleaned up (Klatt 2015).

Grinders manufactured for single dosing are often also designed for a low retention, for example, by eliminating rough surfaces that include hollows or by using magnetic burr mounts that do not require screw holes. Single dosing and low retention are two concepts that go together, because the presence of a hopper typically causes retention. When a barista stops the grinder with a full hopper, some beans will have just been crushed between the burrs and will therefore remain there until the next dose is ground.

I often see misleading experiments in the coffee community, in which baristas weigh a dose of coffee before and after grinding to determine the extent of retention. While this can be useful, it by no means provides a direct measurement of total retention, because some retained grounds can be simply exchanged with fresh grounds, resulting in no apparent discrepancy between the input and output coffee weights. While retention may introduce some variation in the output coffee weight, an accurate measurement of retention would require a deep cleaning and differential measurements (output minus input) across several doses so that the inner parts of the grinder that retain coffee grounds have enough time to gradually fill up and cause a difference between the output and input coffee weights.

3.13 Particle Shapes

One aspect of coffee particles we have not yet discussed in detail is their shape. The shapes of coffee particles are often far from spherical, as is often assumed, and this can have an important effect on both the flow of water and the dynamics of extraction. Elongated particles having the same particle surface as spherical particles will require water to follow a longer path around the particle and will therefore result in a slower flow for the same 10th percentile of particle diam-

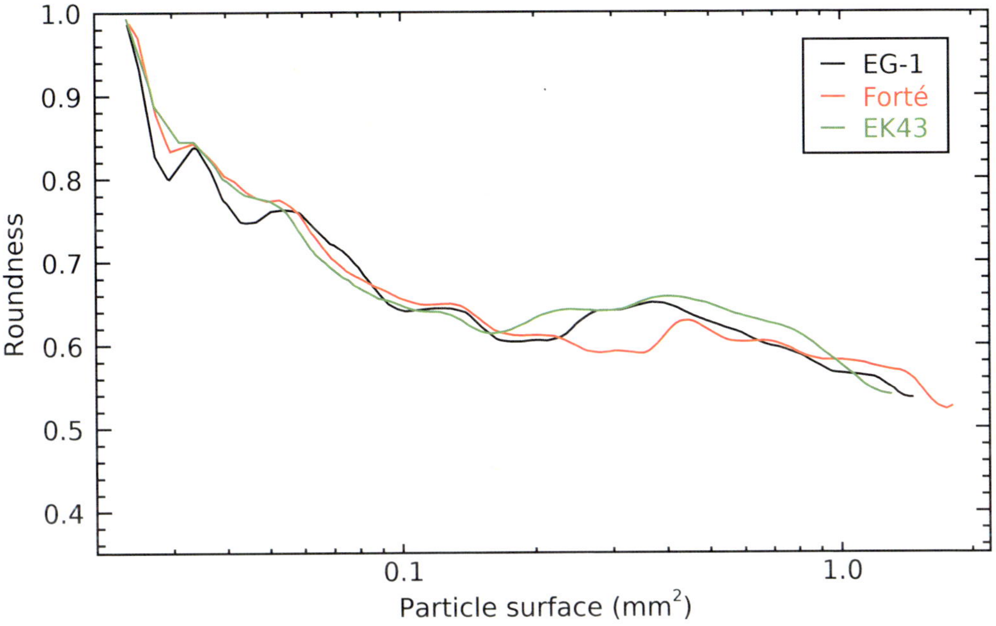

Average particle shapes (darker-colored lines) and representative spreads (lighter-colored lines) for different coffee grinders, as a function of particle surface. The roundness is defined as the surface of a particle divided by the surface of the smallest circle that encompasses the full particle: a roundness of zero means that a particle is shaped like a long stick, whereas 1.0 means a particle appeared as a perfect circle on the scanned image. Coarser particles are systematically less spherical than smaller particles, and the trend does not depend significantly on grinders.

eters D_{10} (Zięba 2017). Elongated particles will also trap a smaller amount of coffee cells deep below their surface, which means the path from coffee solubles to the surface will be reduced and they will contribute flavor profiles more like those of small coffee particles.

Using scanned images and my Grind Size Application, I investigated how much the particles deviate from spherical shapes. I found that the smaller particles seem more spherical than the coarser ones, regardless of the type of coffee grinder used. This may be explained in part by the limited resolution of the scanner blurring the shape of the smallest particles, but the data clearly shows a continuing trend, even among the well-resolved (i.e., much larger than a pixel), coarser particles (with surfaces above 0.1 mm^2). The larger particles are less spherical, consistent with the results of Corrochano et al. (2015).

I found that this particular curve of particle shape versus size never changes significantly, regardless of the coffee roast, origin, variety, or bean temperature. It did not seem to evolve during my seasoning of the EG-1 v2 grinder, nor does it depend on bean temperature, grinding rpm, or small changes in grind size.

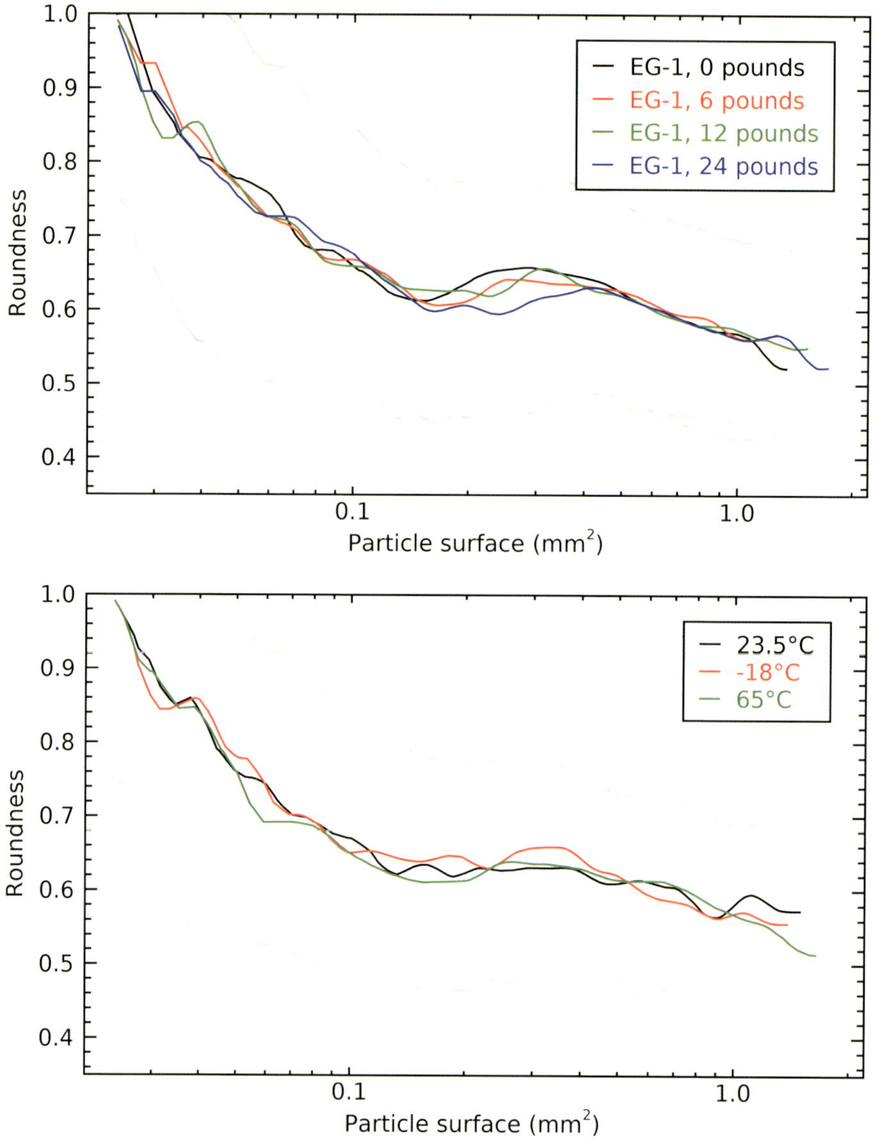

Average particle shapes as a function of grind size for different steps during the seasoning of my SSP ultra-low-fines burrs on the EG-1 v2 and for different bean temperatures, when the burrs were fully seasoned. The trend in particle roundness versus size does not seem to depend strongly on any of these parameters.

CHAPTER 4

Percolation

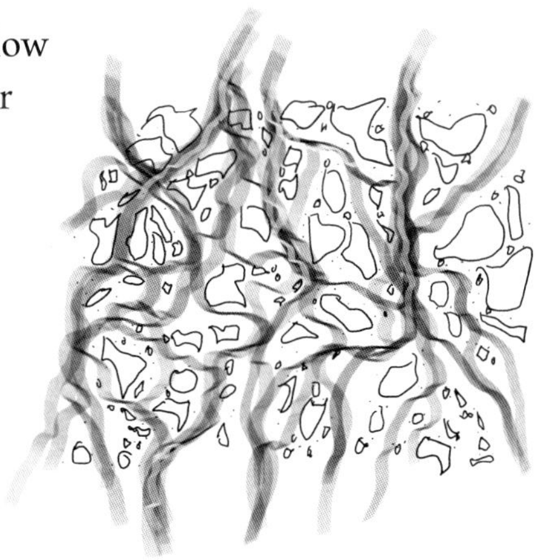

Percolation is the action of letting a fluid flow through a porous material. It happens, for example, as we pour water over coffee or as rainwater filters through the soil. Some of the mathematics developed to describe these phenomena were also found to apply to a wide range of other situations, from the spread of pandemics to network security (e.g., Yang et al. 2013).

The main advantage of percolation in the context of pourover coffee is that it allows an efficient filtration that results in a cleaner cup of coffee. All the while, extraction is also occurring, and fresh water, which extracts more efficiently than concentrated water, is constantly being added from the kettle. In other contexts, such as espresso preparation, its advantage mostly lies in this efficiency of extraction, which allows the preparation of a beverage in mere dozens of seconds rather than in minutes. These advantages come at a cost, however; percolation is a complex phenomenon that can go wrong in more than one way, and it can cause uneven extraction through the coffee bed. This can draw astringent-, bitter-, and harsh-tasting compounds from small, localized regions of the coffee bed while also underextracting large regions of the bed, resulting in a weak and unpleasant beverage.

In this chapter, we examine specifically how the physics of percolation can apply to coffee brewing and what it can teach us about evenness of flow and extraction. We examine filtration, clogging, the importance of preinfusion, and how a coffee bed reacts to taps and vibrations. We will use these concepts to better understand preinfusion and *channeling,* and we will reflect on how we can perform percolation in a more controlled and repeatable way. We will see that the flow of water is a crucial concept that requires our constant attention: maintaining a stable and even flow is one of the main challenges and goals in preparing a good cup of percolated coffee.

4.1 Darcy's Law

Darcy's law is the first step to understanding the physics of percolation. It was first derived directly from observations to describe how a fluid flows through a porous medium such as ground coffee, when placed in a cylindrical container as illustrated at right.

The equation for Darcy's law is the following:

$$Q = \frac{kA}{\mu L}\left(\rho g h - \Delta p\right)$$

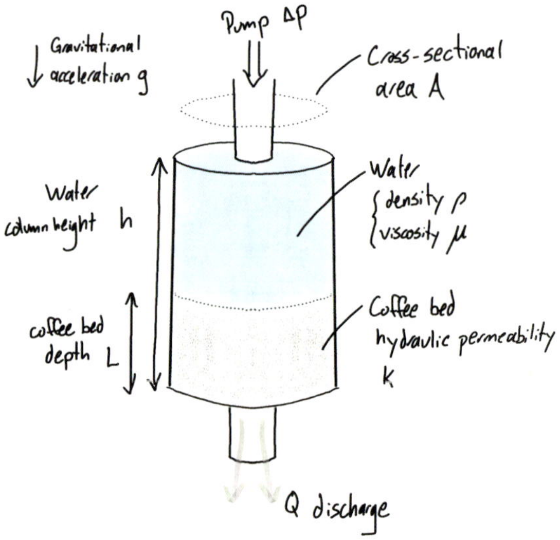

A visual representation of the terms involved in Darcy's law

In order to understand what it means, we first need to define each of the variables represented with letters in the equation above. Q, the quantity Darcy's law allows us to calculate, is called the discharge. It represents the volume of fluid that comes out at the bottom of the cylinder per unit of time. In the context of coffee, this would simply be the volumetric drip rate of coffee out of the dripper, so I will also call it the drip rate. The variable k represents the hydraulic permeability; it is an intrinsic property of the coffee bed that tells you how easily it lets water through. It mostly depends on grind size and how compacted the bed of coffee is. A is the cross-sectional area of the cylindrical dripper; μ is a property of fluids called the viscosity;[21] and L is the depth of the coffee bed.

All the terms between parentheses represent driving forces that push the water downward through the porous medium. The first three terms are ρ, the mass density of water; g, the downward acceleration caused by Earth's gravity; and h, the total height of water in the dripper. When multiplied together, they

[21] This is also called dynamic or absolute viscosity. Viscosity, sometimes referred to as the "thickness" of a fluid, represents its resistance to deformations.

represent the downward pressure caused by water's own weight. The last term, Δp, represents a drop in pressure caused by a pump either pushing above the dripper or sucking below the dripper. My choice of only including mechanical pressure drops in Δp is personal, and others sometimes choose a convention where the terms ρgh are part of its definition, and therefore "hidden". The minus sign in front of Δp, also a matter of convention, ensures that a top-to-bottom drop of pressure increases the discharge Q. In most applications of filter coffee, the last pressure term is Δp zero, because we rely only on gravity to push water through the coffee bed.

It is important to realize that Darcy's law applies at every moment in time; as coffee drips from the container and the water column diminishes, and as the water gets cooler, variables on the right of Darcy's law will change over time, leading to a change of drip rate Q over time.

Basically, Darcy's law tells us that

- A more permeable (i.e., coarser ground, less tamped) coffee bed will drip faster.
- A wider container will drip more coffee per second.
- A more viscous fluid will drip slower.
- A thicker coffee bed will drip slower.
- A denser fluid will drip faster.
- Coffee will drip faster while the column of water is taller.

To give you an idea of the relative scales at play, an 8-cm-tall column of water that is typical of a full V60 corresponds to a downward pressure drop of 0.008 bars. This is about a thousand times less pressure than the typical 9 bars an espresso machine provides, so you can already get a sense of why we must grind much coarser for gravity-driven brews—and make the hydraulic permeability k much larger—to get decent brew times.

In the context of gravity-driven cases ($\Delta p = 0$), the hydraulic permeability is sometimes represented in an alternate form called the *hydraulic conductivity* K, such that

$$K = \frac{k\rho g}{\mu} = \frac{1}{R}$$

The inverse of K is sometimes also called the *hydraulic resistance* R, and it is perhaps an easier quantity to interpret because it represents how much resistance the coffee bed offers to the flow of water. For this reason, I often talk in terms of hydraulic resistance in this book, but any claim about resistance can be reversed and then made to be about hydraulic conductivity. The definition of hydraulic resistance can be used to rewrite Darcy's law in a slightly simpler way, in the context of gravity-driven brews:

$$Q = \frac{Ah}{RL}$$

We can learn something very interesting just by playing with Darcy's law. If we focus on the case of drip coffee and express Darcy's law in terms of the hydraulic permeability k, we can rearrange the equation in the following way:

$$Q = \frac{kAgh}{\nu L}$$

Here, I combined the viscosity μ of water and its density ρ into a new quantity often used in hydraulics that is called the kinematic viscosity ν:

$$\nu = \frac{\mu}{\rho}$$

This is useful because both the viscosity μ and mass density ρ of water depend on its temperature: they both decrease as temperature increases. However, they have opposite effects on the drip rate: decreasing density makes the drip slower because there is less weight pushing downward, and decreasing viscosity makes it faster. Therefore, combining both effects into ν allows us to see which of these two competing effects wins over. There is already plenty of data in the literature about how the kinematic viscosity of water ν changes with temperature.

As we can see from the figure on the opposite page, the decrease in viscosity μ as water becomes hotter completely dominates over the decrease in density ρ, leading to a decrease in kinematic viscosity ν. As a consequence, coffee drips faster when we use hotter water. This is not a small effect, by any means; it is not rare to see baristas brewing in the range of 80°C to 95°C (176°F to 194°F), corresponding to a flow rate 20% faster at the higher end of this range. Darcy's law has already taught us something we can directly apply in real life: using cooler brew water will require us to grind coarser if we want to compensate for the slower flow of water caused by its increased viscosity.

Darcy's law can teach us another practical fact already quite well known in the coffee world: using a smaller dose of coffee will require grinding finer to obtain the same drip rate. The exact way in which the bed depth L changes with coffee dose depends on the dripper geometry; put simply, a conical dripper such as a V60 will make this change much less important with typical coffee doses (>10 g) and is therefore a bit more versatile.

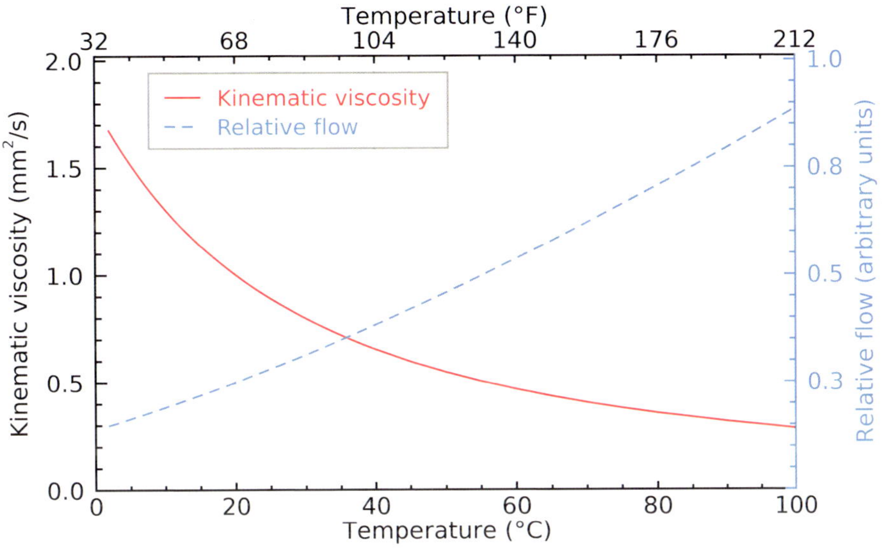

Kinematic viscosity of water as a function of temperature (red line) and its effect on flow according to Darcy's law (dashed blue line)

Darcy's law has several limitations because it relies on a few specific assumptions about the situation we are in:

- **Dripper shape:** Darcy's law assumes a cylindrical dripper. This is the least problematic limitation because a branch of mathematics called differential calculus allows us to adapt Darcy's law to different dripper geometries in a way that I will not detail here. This is done in the Appendix, but for now, it is enough to know that we can generate variants of Darcy's law for different dripper geometries.
- **Turbulent flow:** If the flow of water through the coffee bed becomes so fast as to be turbulent,[22] then the observed drip rate will become slower than what Darcy's law predicts, as if the coffee bed had gained an additional source of hydraulic resistance often called *turbulent friction*. However, we would need a kilometers-tall water column to produce this effect, and therefore it is really not a concern for filter coffee. Even espresso brewing does not typically reach the pressures required to encounter turbulent flow inside the coffee bed (e.g., see Ellero and Navarini 2019). It is important to note that water landing on the coffee bed in a turbulent flow does not mean the flow remains turbulent *inside* the coffee bed; only the latter scenario would cause a slower drip rate.
- **Fines migration:** Darcy's law assumes that the particles in the coffee bed are

[22] This is when flow develops complicated fractal-like patterns, illustrated and discussed further in Chapter 6.

immovable and perfectly stable. In reality, this is not the case: fine coffee particles smaller in size than the average spacing between the larger coffee particles can be dragged with the flow, clogging the pores farther down in the bed of coffee or in the paper filter (e.g., see Yang 2016).

- **Channeling:** Coffee particles of all sizes can be moved around if you pour so fast as to dig a hole in the coffee bed. Even if the coffee bed remains flat, a fast flow of water localized inside the coffee bed can also displace coffee particles and create a channel—a region with larger pores and faster flow. A somewhat analogous problem of deforming pores was studied by Rosti et al. (2020).
- **Bed compaction:** The pressure difference between the top and bottom of the coffee bed—also called the pressure drop—caused by a pump or a tall column of water can compress the coffee bed and temporarily reduce its porosity, which will in turn reduce its hydraulic permeability. This effect seems important for espresso brewing, but I have never observed it clearly when looking at filter coffee data, and I suspect the changes in pressure drop caused by the weight of water are just not significant enough to affect porosity. In contrast with the migration of fines, this effect is reversible in the sense that the porosity will come back to its initial value when the pressure drop is reduced again. This effect explains why the coffee bed suddenly gains resistance when pressing down harder on an AeroPress plunger.
- **Bypass flow:** If water finds a way to *bypass* the coffee bed, the observed drip rate can become much faster than what Darcy's law predicts. As we will see in Chapter 7, this effect is more likely to occur with conical coffee drippers. If we know where the water passes and how much hydraulic conductivity these alternate paths are offering, we can again adapt Darcy's law to rectify predictions of the drip rate.

Some of the phenomena above can still be treated by using Darcy's law if one adapts the hydraulic permeability k of the coffee bed so that it changes over time; channeling and bypass would cause it to increase, whereas fines migration, turbulent flow, and bed compression would cause it to decrease. Some of these changes (e.g., channels and fines migration) are irreversible, in the sense that they will not go away unless some external modification is applied to the coffee bed, such as *swirling* the dripper to reduce channels or applying an upward pressure drop to send the fines back up.

Another difficulty arises in applying Darcy's law to predict the hydraulic resistance of a coffee bed. One could imagine that precisely knowing the particle size distribution of the ground coffee could allow us to predict its hydraulic resistance, but things are not that simple, for two main reasons that we briefly mentioned in Chapter 3. First, the shapes of coffee particles also have an impact on hydraulic resistance. More-elongated particles require that water travels larger distances around them, thereby slowing the flow of water and resulting in a larger hydraulic resistance, even for a fixed particle surface (Zięba 2017). The surface roughness of coffee particles can also be counted as shape irregularities, in the sense that rougher particles will also cause a higher hydraulic resistance.

Second, the particle size distribution informs us on the size of the coffee particles, but the distribution of spacings between the coffee particles is what determines the flow of water; thus, we are required to reverse the particle size distribution into a pore size distribution. This is extremely hard to do in practice, because it depends on how packed the particles are. If the ground coffee is just thrown together loosely, as we do for the preparation of filter coffee, the pores will be large, but the exact same coffee tamped tightly in an espresso portafilter will have smaller pores, on average. In other words, transforming a particle size distribution into hydraulic resistance requires knowing not only the particle shape and surface roughness distribution, but also a detailed knowledge of how packed the bed of coffee is. The mathematics involved can quickly become a nightmare, because most of these quantities are complex and hard to measure or quantify in a precise way. Therefore, either computer simulations or approximate empirical relations are mostly used in hydraulic science, and those can depend on the specific materials in play and the manner in which they were ground.

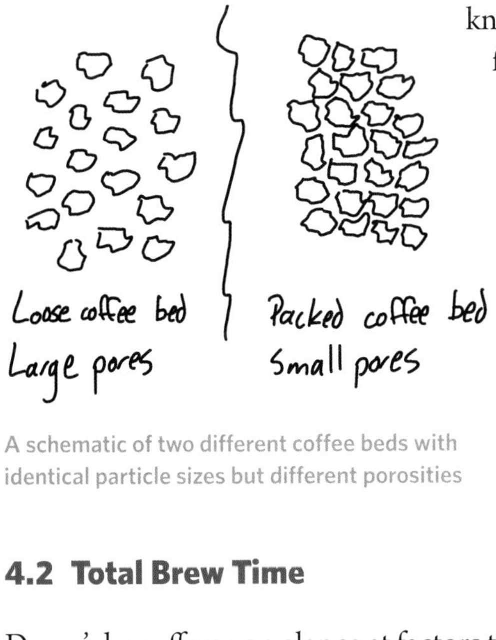

A schematic of two different coffee beds with identical particle sizes but different porosities

4.2 Total Brew Time

Darcy's law offers us a glance at factors that affect brew time because a faster discharge, on average, will obviously result in a shorter brew time. In reality, things are a bit more complicated than this, but we can obtain an intuitive understanding of the most important variables affecting brew time by imagining the following scenario. Visualize a cylindrical dripper into which we pour all the water fast and at once, with very little agitation of the coffee bed, and then wait for the percolation to complete.

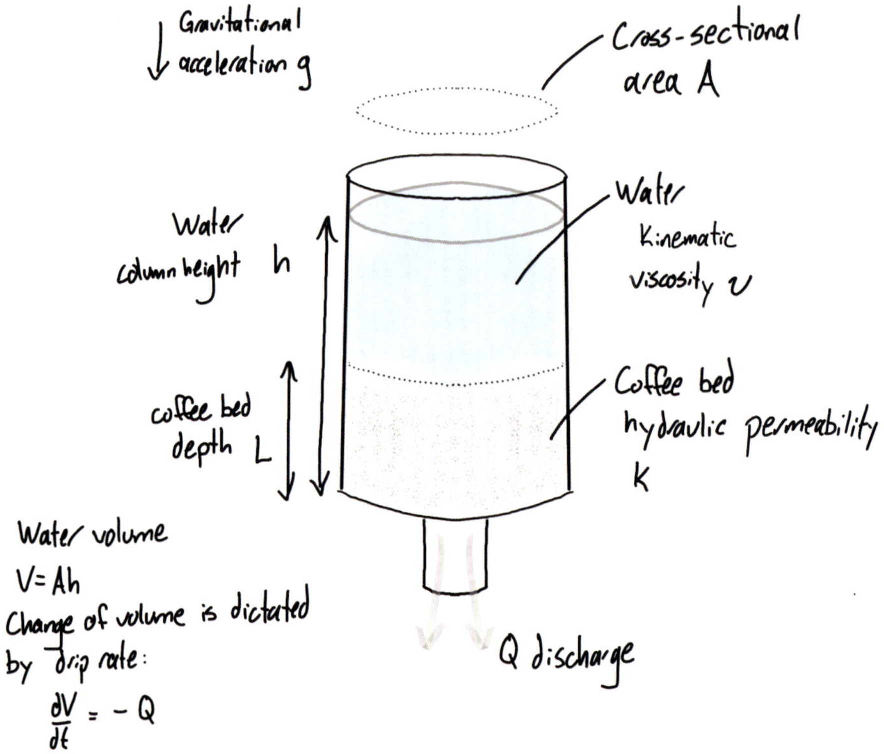

A schematic of a cylindrical dripper with coffee dripping out under the effect of gravity

At the start of the process, water will flow faster because the tall column of water provides a lot of weight to push water through the coffee bed. As coffee flows out, this column will shorten and the drip rate will slow down, until all the water drips out. Solving Darcy's law at every moment in time informs us on how the water column h and drip rate Q evolve in time:[23]

$$h(t) = h_0 \exp\left(-\frac{kg}{\nu L}t\right) \qquad Q(t) = Q_0 \exp\left(-\frac{kg}{\nu L}t\right)$$

Unsurprisingly, both equations contain many terms in common with Darcy's law. Q_0 and h_0 are the drip rate and water column height at the start of the brew, related to each other by Darcy's law, and t is time. Both equations represent a quantity that is decreasing exponentially in time; this is not unexpected, because the drip rate will slow down as the water column height reduces. A graphical representation, shown opposite, can help us understand this more intuitively.

However, we are not done with our calculation of the total brew time, because we have thus far found out how fast the water column shrinks, not when

[23] This calculation is detailed in the Appendix.

it is done dripping completely. We can define the end of the brew as the moment when the column of water is exactly as tall as the coffee bed ($h = L$), corresponding to the moment when we do not see any water above the coffee bed. Solving the above equation for $h(t)$ with this constraint leads to the following equation for the total brew time T:[24]

$$T = L\frac{\nu}{kg}\ln\left(R\frac{\rho_c}{\rho_w}\right)$$

In the equation above, "ln" is the *natural logarithm*, ρ_c and ρ_w are, respectively, the mass densities of coffee and water, and R is the *brew ratio* that was used: for example, 20 grams of dry coffee with 320 grams of total water corresponds to $R = 16$. We can immediately determine a few useful things from the equation above: the total brew time will be longer (larger T) if water is more viscous (i.e., colder), resulting in a larger value for ν. Similarly, grinding the coffee coarser or making the coffee bed more loosely packed would both increase its hydraulic permeability k and make the brew time shorter. Interestingly, brewing on the moon (which has a smaller g) would make the brews much longer—not that this is of any immediate practical use!

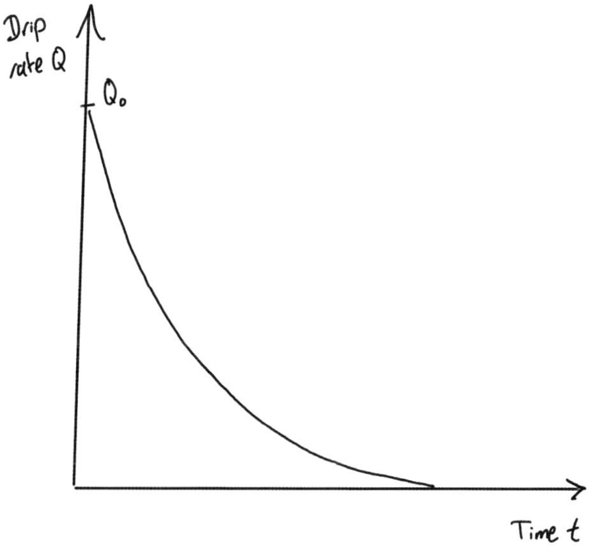

Drip rate as a function of time, as predicted by Darcy's law for a cylindrical dripper

Now let's get to the much more interesting parts of this discussion: the area of the coffee bed A does *not* appear in the equation above! Contrary to intuition, changing the width of a dripper will not affect brew time—that is, if the depth of the coffee bed L is kept the same. This shines a very interesting light on bed depth: it is a truly important variable in brewing that, unlike the weight of the coffee dose, directly controls *drawdown* time. For this reason, I find it more useful to think in terms of bed depth when I reflect on how changing the coffee dose affects a brew. This is consistent with the findings of Rao (2008; 2016) that the depth of a coffee bed is of great importance, although this is not

[24] This is also derived in the Appendix.

the only reason; we will come back to bed depth again later, when we discuss channeling and fines migration. We have already unveiled something interesting about cylindrical drippers: the width of a dripper will dictate what dose of coffee is appropriate for it, because only one dose will produce the desired brew time, with water temperature, grind size, and other parameters fixed.

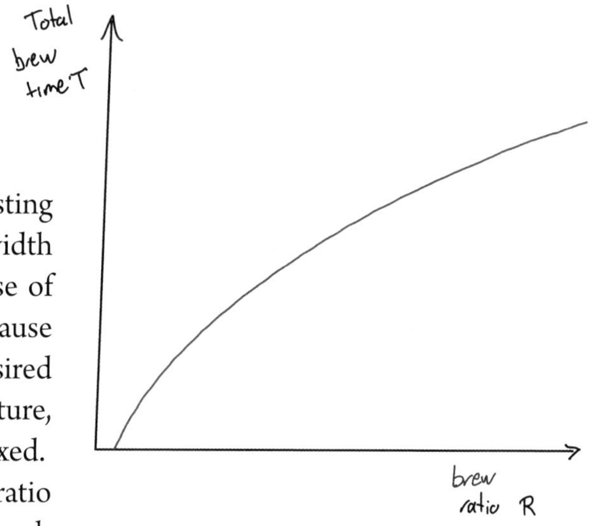

Effect of brew ratio on the total brew time, as predicted by Darcy's law for a cylindrical dripper

Another variable, the brew ratio *R*, plays a role in brew time, although a less important one. The fact that *R* appears inside a logarithm in the previous equation is a way for mathematics to tell us that *T* changes only slowly when we change *R*, as displayed in the graph above.

This can be understood intuitively because adding more water on top of the initial water column will require more time for all the water to drip out, but it will also make the flow initially faster, making the impact of *R* somewhat smaller on the total brew time. It is important to remember we assumed that all water was poured at once into the dripper. If the volume of the dripper limits how much water can be poured at once, the effect of the brew ratio *R* on drawdown time will be more important.

One further complication to the scenario described above is the case of a conical dripper, such as the popular V60. In the Appendix, we derive and discuss an alternate form of Darcy's law that accounts for the conical geometry; the equation is longer and less intuitive, but in reality it is only slightly different from that of the cylinder case discussed above. One of the main differences is that the depth of the coffee bed changes faster for the first few grams of a coffee dose, and then it changes very slowly as the dose increases. This means that conical drippers are a bit more versatile in terms of the range of doses that will produce a given brew time, as long as the dose is not smaller than 10 grams or so (in the case of a V60). The water column in a conical dripper will also change more slowly at first and then later drop precipitously, compared with a cylinder. We can visualize this schematically on the opposite page.

We can see that a conical dripper will have a different evolution of its drip rate compared with a cylindrical dripper. We will come back to this discussion in Chapter 7 when we discuss dripper properties.

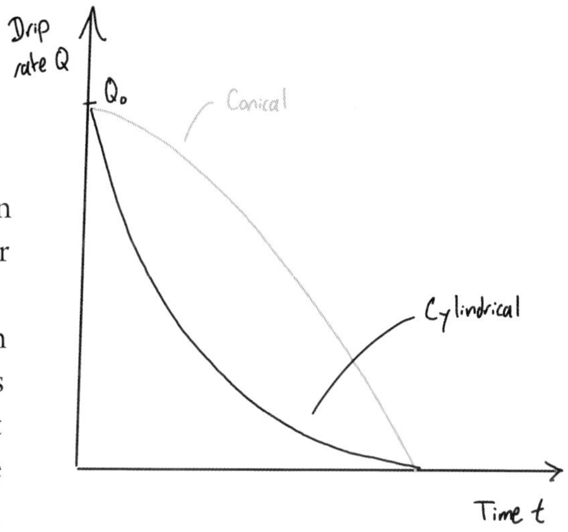

Drip rate as a function of time, as predicted by Darcy's law, for cylindrical and conical drippers

While the mineral composition of brew water has an impact on its viscosity, it is small and will not affect brew time significantly except in the most extreme cases of hard water (in the hundreds of thousands of ppm as $CaCO_3$; Ozdemir et al. 2007). The exact coffee filter that is used will usually have an impact on the total brew time, although the impact is rarely large (barring scenarios where the filter clogs). We will discuss this further in Chapter 5. Other details of dripper geometries, such as ridges along the dripper walls or bottom, can drastically impact brew time. They modulate the area of the filter through which water can flow, and sometimes they cause water to bypass the bed of coffee completely; we will discuss those effects in Chapter 7.

There is one variable that we have not discussed—and it has a dramatic impact on total brew time: how fast, and exactly when water is poured into the dripper. The effects are not hard to deduce; a faster pour rate will increase the water height and shorten brew time, and separating the brew into more pours will instead keep the water height lower on average and make the brews longer. The details of how the kettle is used to pour water in the dripper afford us a lot of control on brew time, but it will also affect the average slurry temperature. It also means that any inconsistency in the pour rate will result in an inconsistent brew time. To me, consistency is therefore one of the most compelling arguments for using a precision kettle. We will discuss more details relevant to kettles in Chapter 6.

At this point, I would like the reader to take a moment and consider how useful it is to compare brew times with other baristas when discussing brew recipes. In particular, pay attention to the fact that a different grinder will play a major role on the particle size and particle shape distributions of the ground coffee. Not only that, but coffee origin, moisture content, and its temperature will change the particle size distribution, even if the grinder remains unchanged. The details of how the coffee is placed in the filter can affect how tightly it is packed, affecting its hydraulic resistance. Other aspects we will discuss later, such as filter clogging, fines migration, and channeling, can also affect the hydraulic resistance of a coffee bed differently from one brew to another.

Different baristas may define the end, or even the start, of their brews in different ways. As if those were not enough, other factors—the details of how quickly water is poured into the dripper, the number of pours, and the exact moment at which the pours were begun—will impact brew time dramatically. That last point cannot be stressed enough, as it may be the largest source of variations among those stated before. Obviously, changing the preinfusion time will also affect the total brew time. Keeping all of this in mind may have you realize how futile a precise comparison of brew times can be, unless they are very loose or performed in an extremely well-controlled environment. Based on my experience, reaching a consistent beverage concentration from one brew to the next down to 0.01% in total dissolved solids is easier than consistently obtaining the same brew time within less than 5 seconds.

4.3 Preinfusion

Preinfusion, sometimes also called the *bloom*, is a crucial step in achieving an even flow and extraction during a percolation brew. As the first drops of water are released on the dry coffee bed, several things begin happening at once. The water initially wets only partial regions of the coffee bed, and the hydraulic resistance goes down drastically in these specific regions. This is true because, like sand, dry coffee is much more resistant to the flow of water compared with wet coffee (e.g., see Yeh, Ye, and Khaleel 2005). This is where the more hardcore mathematics of percolation theory play an important role. They determine exactly when and how paths of wet coffee will connect and allow water to flow all the way through, as dry regions are slowly conquered by water. In this discussion, we will not need to explore the details of this phenomenon (see Stanley et al. 2003 for

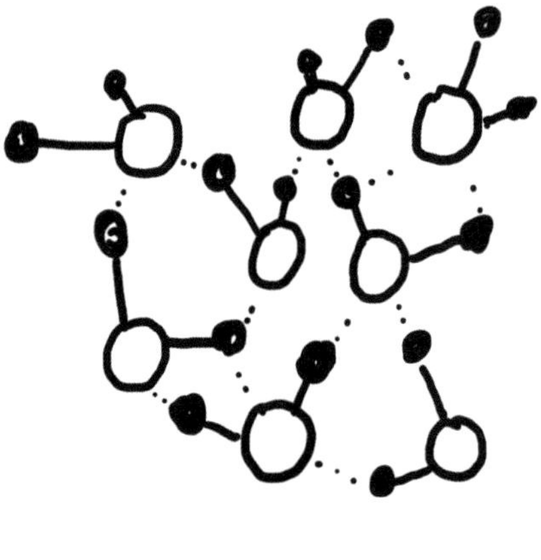

Schematic of water molecules with hydrogen bonds (dotted lines)

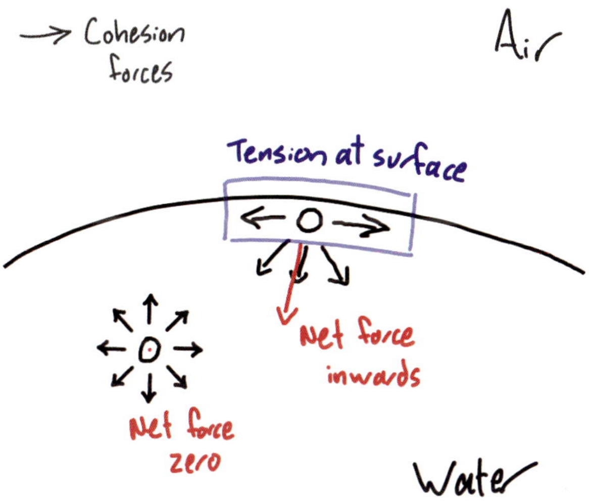

A schematic of surface tension at a water/air interface. Individual molecules of water (black circles) withstand forces of cohesion (black arrows) toward neighbor water molecules, but they have no forces of adhesion to air molecules. As a consequence, cohesive forces cancel out for water molecules well below the surface, and they are affected by a zero-net force, but water molecules near the surface are affected by a net inward force. The opposed and cancelling forces along the surface cause a tension similar to that on a string pulled from both ends.

more details). In practice, we want the grounds to reach full saturation, where all coffee particles become wet, *before* we begin pouring a lot of water and perform the bulk of extraction. If we don't, water will tend to flow preferentially through some paths ("channels"), overextracting these specific parts of the coffee bed while leaving the rest underextracted.

This gradual wetting of the coffee bed happens in part because gravity pulls water downward, encouraging it to find a path of least resistance and travel farther downward. Fortunately for us, another phenomenon, called capillary action, also helps the water propagate through the coffee bed *in every direction.* Capillary action is the phenomenon that causes a towel to progressively get wet farther upward, even if you dip only its bottom tip in water. To understand why this happens, we first need to discuss the concepts of adhesion and cohesion. Cohesion describes the attraction of molecules of water toward other water molecules. We saw in Chapter 2 that water (H_2O) is a polar molecule, with slightly positive charges near its two hydrogen atoms and a slightly negative charge near its oxygen atom. This allows neighboring water molecules to arrange themselves in a way such that opposite charges face each other, causing a force of attraction, forming a hydrogen bond. This intermolecular attraction is the cause for the cohesion of water. Adhesion is a similar phenomenon, but it happens between molecules of water and *another* material. In a sense, cohesion could thus also be called "self-adhesion." In a context where cohesion is much larger than adhesion—for example, near the surface that is exposed to air, as molecules of water are much more attracted to other water molecules than to air—adhesion forces cause the water to behave as if its surface was covered with a stretched elastic membrane. This phenomenon is often called surface tension, and it is a direct

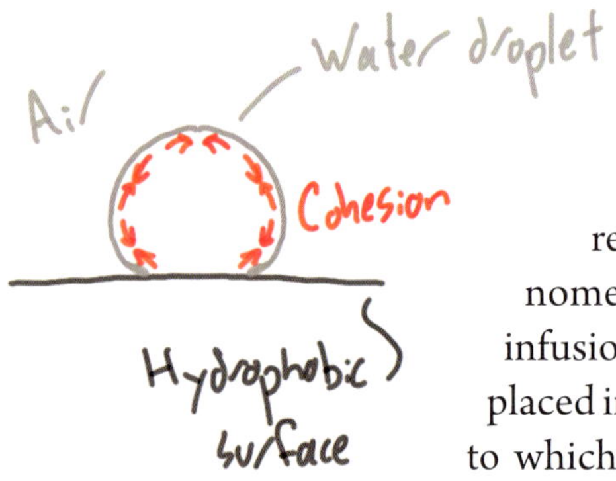

Formation of a water droplet caused by the cohesion of water on a hydrophobic surface, i.e., a surface with very little adhesion to water

consequence of adhesion and cohesion. It is also responsible for the formation of water droplets.

Adhesion and cohesion are also responsible for capillary action, the phenomenon that plays a crucial role during preinfusion. When water (or any other liquid) is placed in a small tube with walls made of a material to which it adheres, the combination of cohesion and adhesion will slowly pull water farther in the tube. This can even cause water to travel against the force of gravity, provided that adhesive and cohesive forces are strong enough.

Capillary action will cause water to begin propagating across the coffee bed, even horizontally, and it will help to even out the initial channels caused by the dry pockets until they disappear completely. Even more importantly, capillary action is what also causes water to enter the particles of coffee—a crucial step that, as we saw in Chapter 1, is needed to dissolve and transport coffee solubles to the surface of the coffee particles. If that did not happen, it would be almost impossible to extract coffee efficiently from intact coffee cells!

While water enters the coffee cells from a series of pores, carbon dioxide (CO_2) stored in the coffee cells during roasting will be pushed away through other pores. This degassing will cause bubbles to appear in the slurry every time water touches a dry particle of coffee. Note that water with more total alkalinity (see Chapter 2) will generate more bubbles during the preinfusion: this is caused by the bicarbonate reacting with acid compounds in the coffee, one product of which is more carbon dioxide. This alone could explain the findings of Hoffmann (2019, Jul. 22) that harder

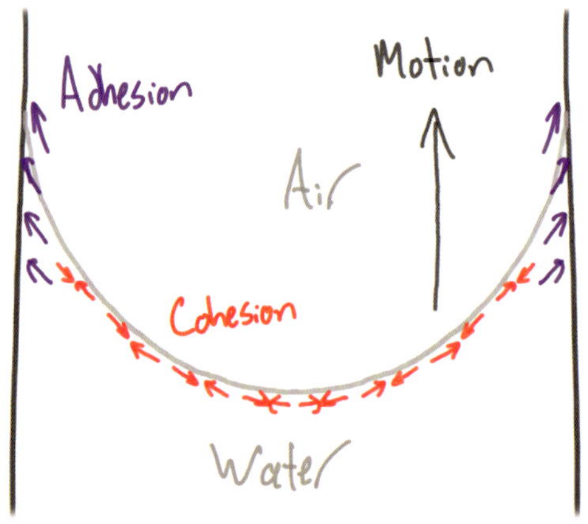

Capillary action, where water moves upward in a tube. The adhesion of water with the edges provides an upward force, causing the water to be lifted near the edges. The cohesion of water lifts up the central part of the water surface.

water produced more bubbles;[25] in other words, it may not have improved the efficiency of the preinfusion, but it might have caused the creation of additional CO_2. Darker-roasted beans, faster-roasted beans, and fresher beans tend to contain more carbon dioxide stored inside the coffee cells under pressure (e.g., Smrke et al. 2018), which will increase the amount of degassing during the preinfusion phase and may, as a consequence, require a more thorough agitation during preinfusion.

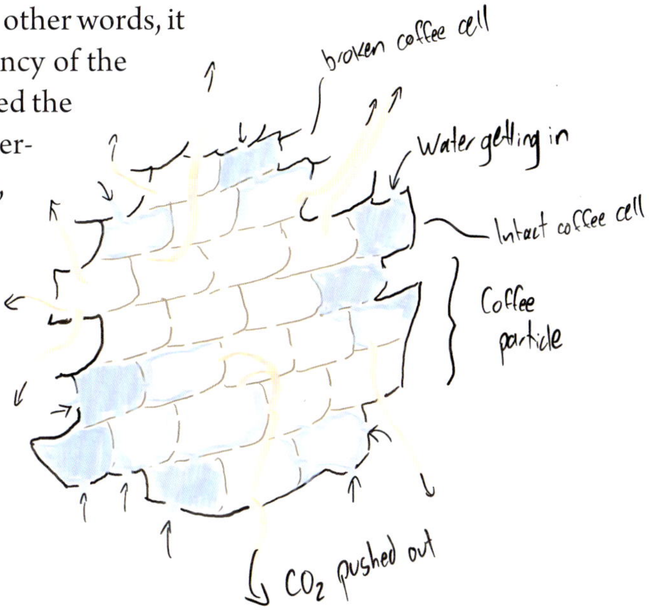

A schematic of water getting into a coffee particle during preinfusion by capillary action and forcing out the CO_2 stored inside the coffee cells

A good preinfusion phase will get much of the coffee bed wet before the next pour and will result in the production of no further bubbles after it is completed. If you notice a lot of large bubbles occurring after the preinfusion, it means you could probably improve your brew by making your preinfusion more effective. The appearance of carbon dioxide bubbles after preinfusion is a problem, not only because they indicate some part of the coffee bed was still dry and will end up lagging behind the rest in terms of its extraction yield, but also because bubbles flowing upward will cause localized channels of much lower hydraulic resistance (Hunt and Manga 2003) through which water will flow preferentially, causing further unevenness of extraction.

It is interesting to note that preinfusion offers no advantage for immersion brews. In such a scenario, the whole brew can be viewed as a preinfusion. An initial short period of agitation might be useful, however, as it will allow all coffee particles to begin extracting at the same time. In most scenarios of immersion, I would recommend agitating in a back-and-forth motion with a wide object (e.g., a spoon or paddle) because this creates a greater amount of turbulence over

[25] Harder water, with higher mineral content and therefore a higher total hardness, tends to also have a higher total alkalinity.

alternative methods such as rotations, and turbulence is extremely efficient at breaking up even the smallest clumps of dry coffee.

The speed at which capillary action allows water to travel through the coffee bed during preinfusion is determined by the Lucas–Washburn equation (Hanžič, Kosec, and Anželc 2010):

$$l = \kappa\sqrt{t}$$

where l is the distance traveled by water at time t, and κ is the capillary coefficient of the porous medium. As can be expected from our previous discussion, this capillary coefficient depends on the average pore size of the medium r and the surface tension γ and the viscosity μ of water (Yang et al. 2019):

$$\kappa = \sqrt{\frac{r\gamma}{2\mu}}$$

This propagation by capillary action can be visualized with the schematic shown below.

The spacings between coffee particles constitute much larger pores than those connecting coffee cells inside the coffee particles for any reasonable filter grind size. Consequently, the propagation of water through the dry coffee bed will happen at a much faster rate than the propagation of water *inside* the coffee particles. Coffee particles will typically require a few minutes before they are completely filled with water down to their cores, especially for coarser grind sizes, because more distance must be traveled to reach their core. For lighter roasts, the pores separating their cells are smaller (e.g., see Oliveros et al. 2017 and Schenker et al. 2000), also slowing down the propagation of water. This slow absorption of water by the coffee particles can be illustrated with the brew data shown opposite, obtained with a method that I will describe in Chapter 11.

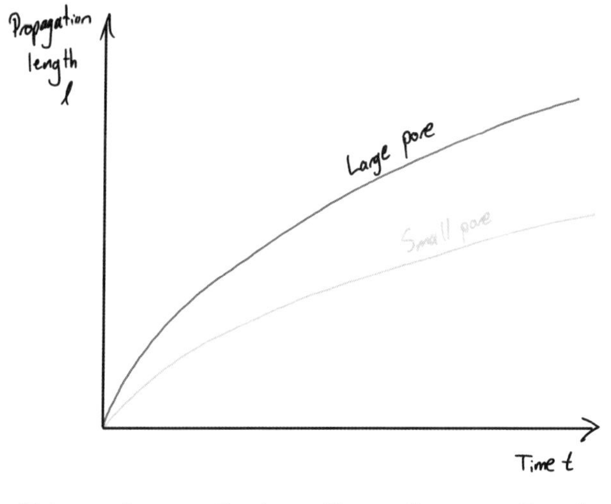

Distance of propagation by capillary action versus time, for two different pore sizes

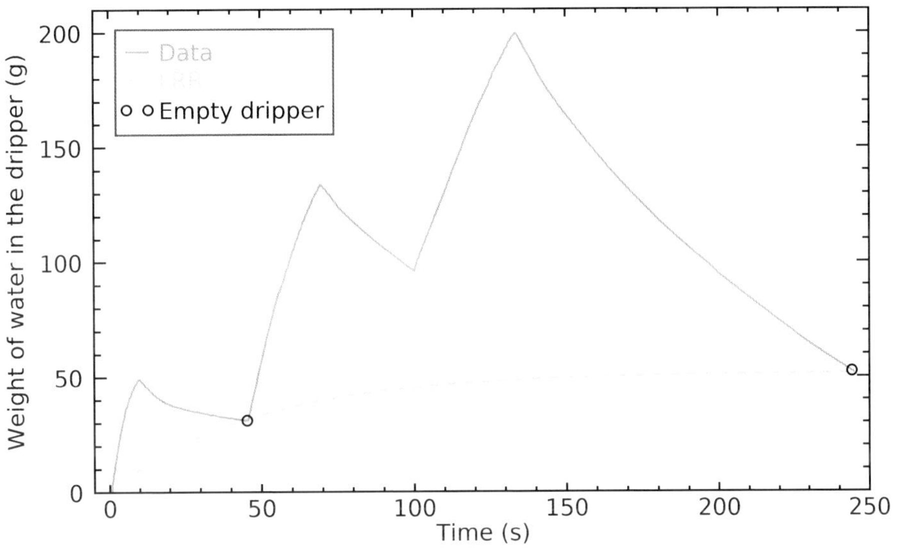

Weight of water in a V60 dripper, as a function of time during a representative brew. The water level dipped below the coffee bed at only two moments (circles), indicating the weight of water retained by the coffee bed at only these moments. An asymptotic model of LRR growth is indicated with dashed lines.

As we see in the figure above, the final liquid retained ratio (LRR)[26] is only reached after about 2 minutes, and the amount of water retained inside the coffee bed can be observed only for short moments between pours, when no water is left above the coffee bed. Experimental measurements allowed me to determine that the rate at which the particles absorb water depends strongly on particle sizes (it is *slower* for coarser particles), but the final LRR did not depend significantly on grind size.[27]

I suspect this effect is what led some baristas to find that 2- to 5-minute blooms can improve the quality of some coffees (e.g., Motoyoshi 2019), although this does not seem like a perfect solution: after water has dripped through the coffee bed during the first few seconds of the preinfusion phase, only the water that was retained between the coffee particles is potentially available for being absorbed, and it is unclear whether this lack of additional water slows down the process of absorption, especially with the lack of high-quality data available on absorption rates. However, in my experience, using 2- to 3-minute preinfusions has increased average extraction yields by as much as 1% compared with

[26] defined in Chapter 1 as the ratio of total water weight retained in the coffee bed divided by the weight of the dry coffee dose

[27] This may only be true of my specific grinder and burrs, an EG-1 v2 with SSP ultra-low-fines burrs.

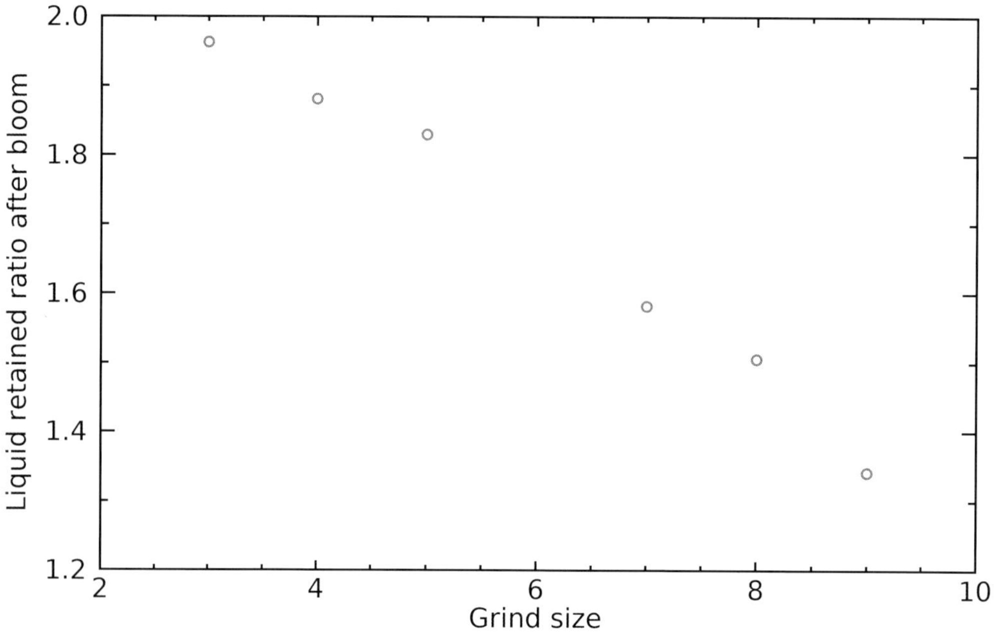

Liquid retained ratio immediately at the end of a 45-second bloom, as a function of grind size. All brews had a final LRR of 2.6 ± 0.1, regardless of grind size.

more standard 45-second preinfusions for several light-roasted coffees, as well as improving their taste.

In the case of darker roasts, I have not observed an improvement in extraction yield, and long blooms resulted in bitter and harsh brews. This could be explained by darker roasts having larger intercellular pores and not requiring as much time to fully absorb water. In that scenario, the longer preinfusion may have extracted some of the less desirable slow-diffusing compounds without any modification to the grind size, slurry temperature, brew time, and other brew parameters.

I suspect it is desirable to achieve a full absorption of water by all coffee particles during the preinfusion because this will allow all coffee cells, down to the cores of the largest coffee particles, to contribute to extraction as fast as possible. If the main percolation phase is started at 45 seconds and absorption completes only after 2 minutes, a large fraction of core coffee cells will have been underextracted, will likely contribute less desirable flavor profiles to the brew, and will waste more coffee solubles.

Measuring absorption precisely is especially hard in practice because it requires refilling the interstitial spaces between coffee particles with slow kettle pours and then waiting for a long time for any extra water to drip. I can hypothesize about what technique might improve absorption during the preinfusion of filter coffee, but more research is definitely needed in this area, and it is far from guaranteed that doing so will always improve taste. I suspect that an initial preinfusion phase with a brew ratio of about 2:1, followed by a longer phase of very slow, pulsed pours might help achieve more complete absorption of water by all coffee particles before the main percolation phase.

This is relatively hard to perform without using a brew ratio of above 3:1 to 4:1 during the full bloom phase, which is not ideal because this means we are left with less water to perform the main extraction phase. In my limited experience, this technique has had a positive impact on extraction yield and taste only for very light-roasted coffee. Using an on/off flow switch that permits a longer bloom without flow would be another solution to properly fill up all coffee particles with water, but I have found it challenging to avoid filter clogging with this method. This situation could possibly be explained by buoyancy lifting the large coffee particles in the absence of flow, allowing fines to migrate down and clog the filter, although more experimentation is still needed.

Speeding up the capillary action of water by changes in water composition is impractical because increasing the capillary coefficient would require extremely hard water (approximately 200,000 ppm as $CaCO_3$ total calcium hardness for a 10% increase in surface tension; Michaud 2016), which would be problematic for many reasons. Changes in carbon dioxide content inside the coffee beans should have only a negligible effect, as the water-to-CO_2 surface tension is not very different from that of water-to-air surface tension at atmospheric pressure and coffee slurry temperatures (see Georgiadis et al. 2010; Nielsen, Bourg, and Sposito 2012; Vargaftik, Volkov, and Voljak 1983). However, CO_2 stored under pressure in the coffee particles could slow down the absorption of water. Increasing the absolute pressure in a sealed bloom would likely accelerate absorption, based on Darcy's law applied to water penetrating the pores inside the coffee cells, but it would be technically challenging and probably a very efficient way to cause explosive messes.

Hotter water has decreased viscosity and slightly decreased surface tension, the combined effect of which is a larger capillary coefficient. Using water as hot as possible therefore seems desirable to accelerate the process of absorption. A change of slurry temperature from 75°C to 88°C (167°F to 190°F) during preinfusion might be expected to accelerate the rate of absorption by as much as 15%, based on the Lucas–Washburn equation discussed above. Preheating the dry coffee dose may be a solution to increase the slurry temperature during bloom, but it could lead to a precipitous loss of coffee aroma.

Believe it or not, yet another phenomenon is taking place during preinfu-

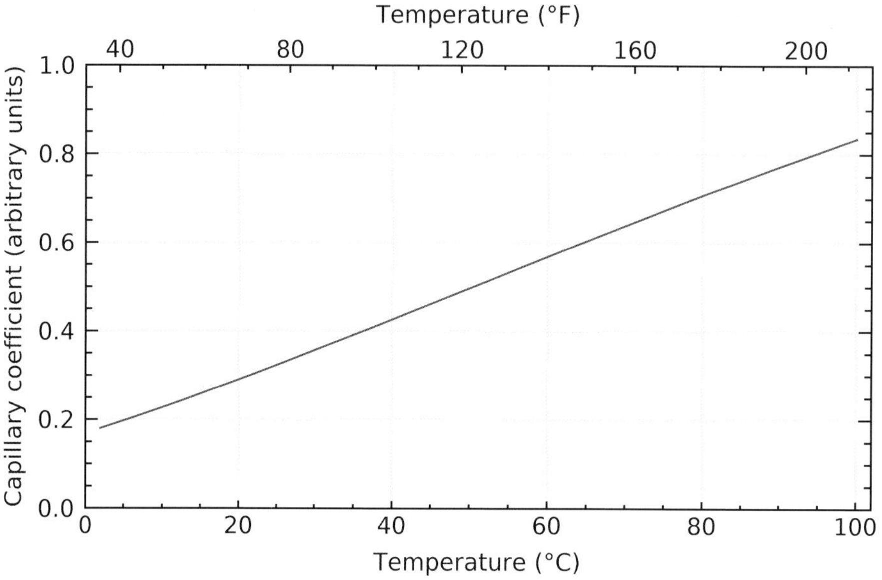

Capillary coefficient as a function of water temperature. The surface tension of water is decreased slightly at warmer temperatures, but its viscosity is decreased dramatically, resulting in a higher capillary coefficient and therefore a more efficient propagation through porous media via capillary action.

sion: the swelling of coffee particles. As the particles of coffee absorb water, they swell up and can cause the coffee bed to look rounder, or "dome-shaped," at the end of the bloom. This effect will be more obvious with larger coffee doses, for purely geometric reasons,[28] but I suspect it may also provide a way of visually estimating how fast the coffee particles have completed their water absorption. This swelling up will likely reduce the porosity of the coffee bed and therefore slightly increase its hydraulic resistance (Aksu, Bazilevskaya, and Karpyn 2015). In some cases, it may also cause the coffee bed to become muddier and easily deformable by the kettle stream. Pouring uniformly across the coffee bed with enough agitation and executing a swirl after the pour (described below) are good strategies to ensure that no channels occur in this scenario. The increased hydraulic resistance may cause brew times to be slightly longer; this can be counteracted with a coarser grind size, faster or fewer pours, if desired.

Creating an initial "nest shape" in the coffee bed has been widely adopted by the specialty coffee community as a practice that improves the efficiency of preinfusion in cone-shaped drippers. Water is first poured into the center of

[28] In a conical dripper, more layers of coffee are swelling up near the center when the coffee bed is deeper.

the nest to allow a faster wetting of the bottom of the coffee bed, and then it is poured around the edges to evenly cover the surface of the coffee bed. This technique does improve the preinfusion phase, in my experience, because it is generally hard to properly wet the bottom of the coffee dose before too much of the water has dripped out and capillary action has more or less stopped its propagation. I suspect this technique would not help much for drippers with flow switches (assuming flow is stopped completely during bloom). Contrary to common intuition, I think the nest shape can even be useful when using flat-bottom drippers because it might also allow prewetting the bottom layers of the coffee bed more quickly. Carving out the nest shape in a dry coffee bed in a repeatable way is not particularly easy; rotating a chopstick near the center of the dry coffee bed can achieve this, although it introduces an undesirable risk of channels because it can easily cause the coffee bed to become too shallow near the center. Malherbe (2020) proposed a method wherein the dry coffee is released over the back of a bent spoon to result in a nest shape without further manipulation. In my experience, simply using a finger with a rotating motion to dig a nest shape has resulted in consistent brew results.

Swirling the coffee bed during the preinfusion phase by holding the dripper base and moving it in small circles was first introduced by James Hoffmann and further popularized by Rao (2008; 2010). The method forces water to move outward horizontally and speeds up the wetting of dry regions, and it destroys channels by rearranging the positions of particles in the coffee bed. This also ensures that the coffee bed is flat before the main percolation phase, especially if the dry coffee was positioned in a nest shape or had swollen, causing a dome shape. Because swirling can destroy channels, it is not only useful during preinfusion, but also later, during the main percolation stage, as a way to improve the evenness of flow. Using a laboratory orbital shaker makes it possible to perform precise and repeatable swirls, but leaving it swirling for a full brew does not seem to improve the results and may favor the bypass of water around the coffee bed or overextraction of the sides of the coffee dose.

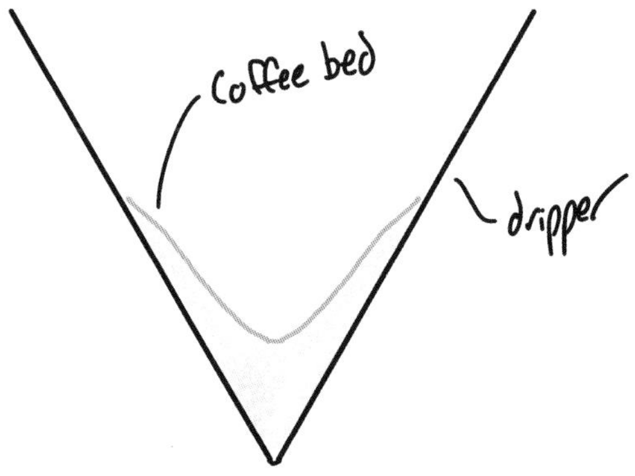

The nest shape is useful to rapidly get the bottom of a coffee dose wet during preinfusion.

4.4 Fines Migration

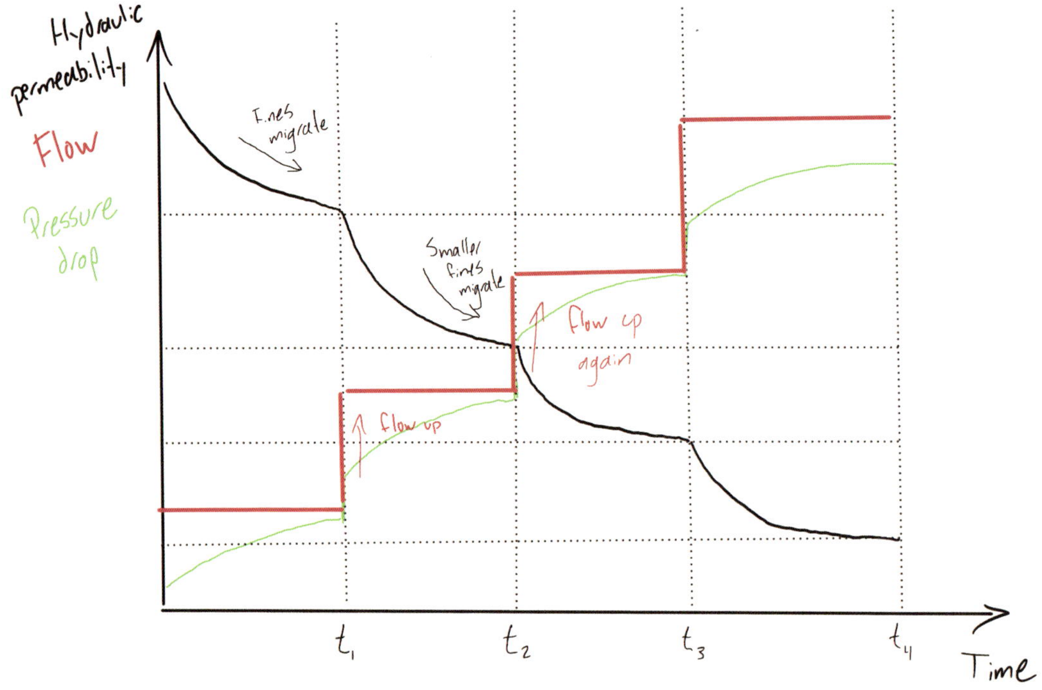

Temporal evolution of hydraulic permeability (black line) in a porous medium containing fines. A pump pressure (green line) is adjusted to maintain constant flow (red line) and then ramped up suddenly to increase flow, in four distinct occurrences. The first segment begins at a low pressure, which induces a relatively slow, localized flow of water between the particles of the porous medium, dragging only the largest fines along with the current. Fines get captured farther down in the porous medium, causing a decrease in hydraulic permeability, which requires pressure to slowly ramp up to maintain flow. The hydraulic permeability stabilizes to a constant value, as no more fines are migrating. When flow is increased by suddenly ramping up the pump pressure even more, the hydraulic permeability begins decreasing again because the faster localized flow of water applies a lever strong enough to dislodge yet smaller fines, until those too are depleted and the hydraulic permeability again stabilizes. Every increase in flow causes a similar event of yet smaller fines migrating and further reducing the hydraulic permeability.

Fines migration is a well-studied phenomenon in the context of damage to soil or filter permeability (e.g., Moffat 2002) but is rarely discussed in the context of coffee brewing. Petracco and Liverani (1993) established that it was the most likely cause for an observation that the puck resistance in espresso preparation gradually increased for several seconds at a fixed pressure, and Cappuccio and Liverani (1999) observed that the reduction in flow was more significant when a larger pump pressure was applied.

These observations are in line with the geophysics literature on the subject

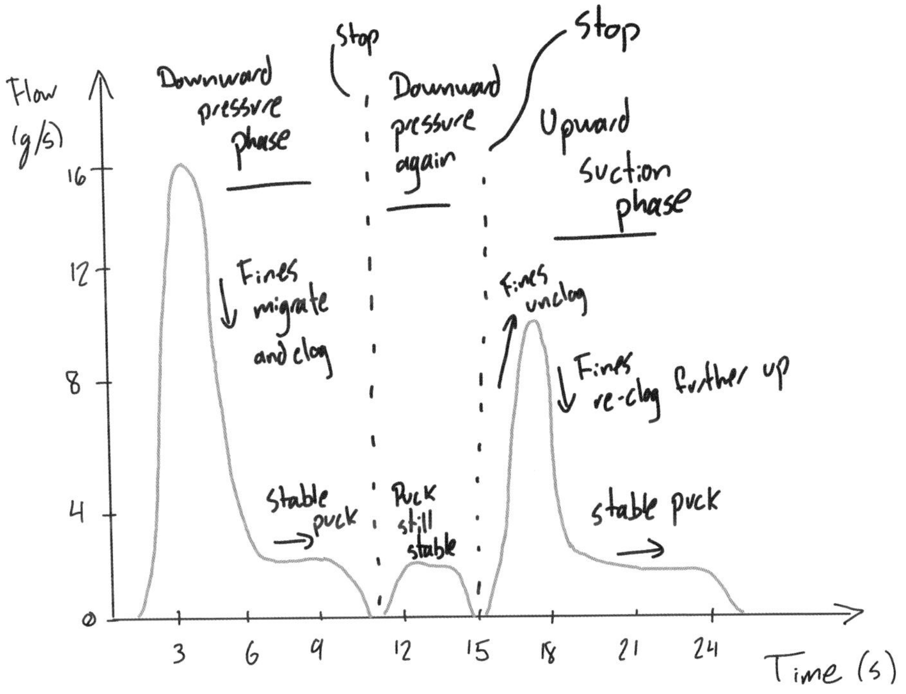

The experiment of Petracco and Liverani (1993) demonstrated that the decrease in hydraulic permeability can be partly reversed by reversing the pump's direction, providing evidence that fines migration and recapture can explain the observed change in permeability.

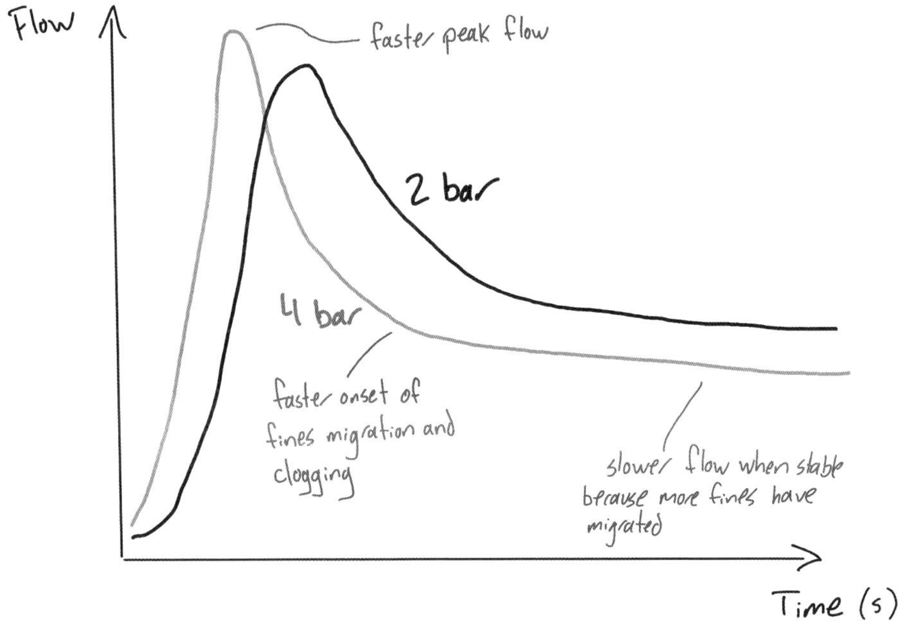

Flow as a function of time for two different pump pressures (Cappuccio and Liverani 1999). A higher pump pressure caused more fines to migrate which increased the puck resistance, resulting in a slower stable flow. Compaction of the coffee bed caused by the larger pressure drop will also contribute to increase its resistance, in addition to fines migration.

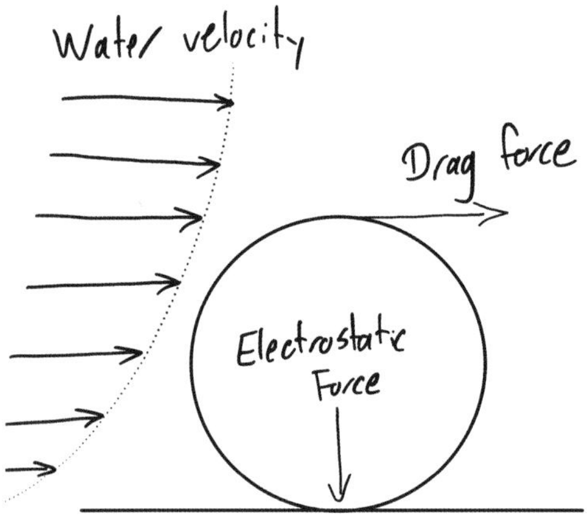

Coffee fine (circle) held to the surface of a larger particle (horizontal line) by electrostatic forces. The flow of water (arrows at left) applies a drag force on the particle, acting as a lever that may dislodge the fine.

of fines migration. For example, Ochi and Vernoux (1998) demonstrated that fines migration causes a decrease and then stabilization in hydraulic permeability at a fixed flow of water through a soil bed, as all fines get displaced and subsequently recaptured deeper in the percolation medium. When the pump pressure is then increased to force a faster flow, a new class of fines begins migrating and causes yet another event of reduction, followed by stabilization in hydraulic permeability. This is illustrated in the graph shown at the beginning of this section.

A faster flow of water induces the migration of additional fines for a simple, yet counterintuitive, reason. A key point is that not all fines can always be moved by the flow water. When a brew is started, fines are usually stuck within the crevices of the larger, irregularly shaped coffee particles or adhere to their surfaces by electrostatic forces. The flow of water will push these fines and try to drag them along, but it must overcome this electrostatic force to do so. A schematic of the situation, shown above, will help us understand how this happens.

The drag force caused by the flow of water acts like a lever on the coffee fine, one side of which is stuck to the surface of a larger particle. The larger the fine, the better the lever will be, and the more it will be susceptible to be dragged away with the flow—provided it is small enough to pass through most of the spacings between the larger coffee particles and therefore qualify as a coffee "fine." Another way to say the same thing is that *the smallest fines are better at hiding from and resisting the flow of water*. The equation for the drag force F_d of a fine of radius r is given by

$$F_{\rm d} = u\mu r$$

where u is the flow of water at the center of the pore and μ is the viscosity of water.

In the hydraulic permeability schematic we discussed before, the first stage, with a slower flow of water, allowed dragging only the largest fines with it. As the velocity of water increased, additional smaller fines became available for migration, which caused the further degradation in hydraulic permeability.

Interestingly, Bergendahl and Grasso (2003) noted that increasing the water temperature will favor a decrease in electrostatic forces between fines and larger particles. This could mean that hotter brew temperatures increase fines migration, but experimental data in the context of coffee brewing will be needed to determine how important this effect is.

One thing we can already deduce from the drag force equation is that local regions where channels form will cause more fines to migrate with the flow of water because it is much faster within a channel.

The second important point in understanding fines migration is what happens to the fines that got dragged away with the flow of water. There are two possible outcomes, in the context of coffee preparation: fines can get stuck (the technical term is "strained") in a pore too small for them to fit through (e.g., Yang 2016), or they can pass through all the way to the coffee beverage. The first case can happen either farther down within the coffee bed (i.e., the fine gets trapped between two other coffee particles) or in a paper filter (i.e., the fine gets trapped in a filter pore that is too small for it to pass through). In both cases, the hydraulic permeability of the system will be reduced, as observed by Ochi and Vernoux (1998). In the latter scenario, where a coffee fine makes it all the way to the beverage, the result will be a cloudier brew that will have more body and mouthfeel but less clarity.

At this point, it is important to realize that the local flow u, which causes fines to migrate, is a different concept from that of the discharge Q, which we discussed in the context of Darcy's law. The discharge, or drip rate, represents the total volume of water that drips in a fixed interval of time, whereas the local flow represents how fast the water is moving through the coffee bed. We can relate the two quantities u and Q with the following equation:

$$u = \frac{Q}{A\varphi}$$

Here, A is still the cross-sectional area of the dripper, and φ, the Greek letter phi, represents the porosity of the coffee bed, sometimes also called the *void fraction*. The porosity is simply defined as the ratio of the empty volume to the total volume before water is poured into the dripper. The more packed a bed of coffee, the smaller the porosity is because there is less empty volume by fraction.

To help us interpret the equation above, imagine we had a cylinder with water flowing in it but no coffee. This corresponds to a 100% porosity, or $\varphi = 1$, because there is only empty space in the cylinder before water begins flowing. In that context, it may be easier to see why multiplying the average flow of water

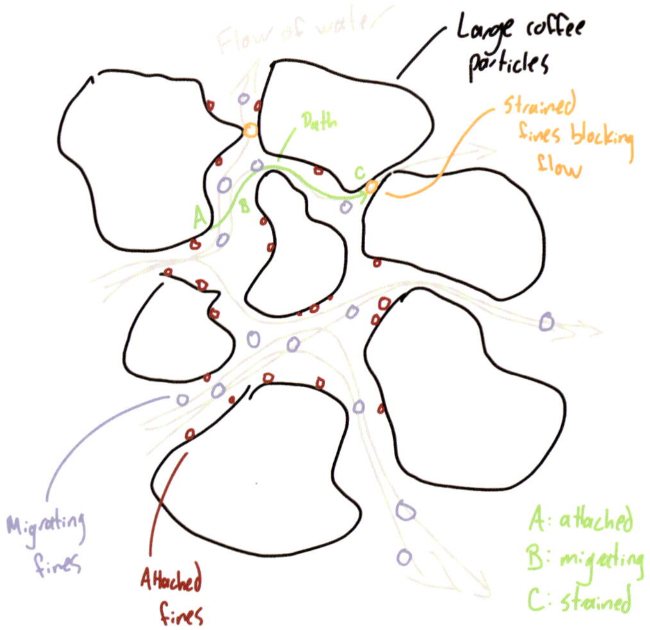

A schematic of fines at different stages during migration. The smallest fines (red) remain attached to larger coffee particles (black) because the flow of water (gray arrows) is not strong enough to dislodge them. Larger fines get detached (purple) and migrate with the flow; some of them get stuck in a pore too small for their size (orange), blocking flow and reducing overall hydraulic permeability. A typical path taken by a fine is shown in green, with its three stages of evolution (A through C).

u (in meters/second) with the cross-sectional area A (in meters^2) gives us the discharge Q (in meters^3/second), as if every point at the surface acted like a small hose contributing water with a flow u to the total discharge Q. When we have coffee in the cylinder, however, our porosity φ will be between zero and 1, and to get the same discharge, each element of water will need to move slightly faster because the effective cross-sectional area A that contributes to the total discharge Q is smaller.

We can now go back to Darcy's equation, discussed earlier, and use the equation above to express the local flow u as a function of other variables, while remembering that a larger flow u will free up a more important population of fines that can be dislodged:

$$u = \frac{k}{\varphi \mu L}\left(\rho g h - \Delta p\right)$$

Now we can see what variables affect the local flow; however, this equation is made a little more obscure by the presence of both the hydraulic conductivity k and porosity φ. If we had a coffee bed that was more porous, for example, both quantities would increase, but it remains unclear how they would affect the local flow u because they are on either side of a ratio and therefore compete for their effect on the local flow. We thus need to dig a big deeper to find out which one "wins" and affects local flow the most.

As mentioned earlier, the hydraulic permeability k depends on grind size, bed packing, and even the shapes of coffee particles. As we discussed in Chapter 3, scientists typically resort to experimental approximations or computer simulations to calculate k, because doing so from fundamental principles is impractical. We already saw that the hydraulic permeability depends on the square of the "10th percentile" D_{10} of a particle size distribution (Anderson et al. 2007; Hazen 1911):

$$k = f_{\mathrm{sp}}\left(D_{10}\right)^2$$

The hydraulic permeability is therefore obtained from the product of the square of D_{10} with an unknown quantity that depends on particle shapes and packing; here we called this f_{sp}. If we kept particle shapes and packing fixed and changed their sizes only, the equation above would allow us to predict changes in hydraulic permeability.

The second useful approximation can also be gleaned from experimental data. A large variety of simple equations have been used to represent the observed variation of permeability as a function of porosity φ (e.g., see a review by Hommel, Coltman, and Class 2018), but one relation that seems useful in a wide range of scenarios is based on a *power law*:

$$k = f_{\mathrm{sd}}\varphi^{\eta}$$

Here, f_{sd} is a new quantity that depends on the shapes and diameters of particles, and η, the Greek letter eta, represents the intrinsic properties of the porous material. Depending on the material, the value of η can range from 2 to well above 20 (Bernabé and Evans 2003), where larger values represent more drastic changes in hydraulic conductivity for the same change in porosity. The most commonly used value across materials is $\eta = 3$ (Bear 1988). We can combine the equations above into a new equation that specifies the respective contributions of porosity and particle sizes to the resulting hydraulic conductivity:

$$k = f_{\mathrm{s}}\left(D_{10}\right)^2 \varphi^{\eta}$$

We are therefore left with an unknown quantity f_s that depends on the shape distribution of coffee particles, but now we see more clearly how the hydraulic conductivity changes with respect to particle size (through D_{10}) and porosity (through φ^{η}). I know this has been a big detour, but bear with me. We can now use this expression of k and insert it back into our modified Darcy's equation to obtain

$$u = \frac{f_{\mathrm{s}} D_{10}^2 \varphi^{\eta-1}}{\mu L}\left(\rho g h - \Delta p\right)$$

This new equation might seem ugly or complicated, but it contains some powerful conceptual insights: it tells us in an unambiguous way which parameters make fines migration less or more of a problem. Because the power η is always larger than 2 (Aharonov, Tenthorey, and Scholz 1998; Colón, Oelkers, and Schott 2004), it already tells us that a larger value of φ increases k in the numerator more than it increases φ in the denominator. This equation tells us other things as well:

- A larger pressure differential between the top and bottom of the coffee bed (pressure drop) will generally increase fines migration. This is true whether the pressure drop is mechanical (Δp here) or due to a very tall column of water (h). However, remember that pump-driven pressure drops are often hundreds if not thousands of times stronger than gravity-driven flow.
- A more porous (φ) coffee bed will be more susceptible to fines migration. Realistic values for the power term η are always above 2, and therefore the power η–1 remains positive and the local flow dependency on porosity never reverses. Compressing or tamping a coffee bed will reduce its porosity φ and therefore also reduce fines migration, assuming other variables in the equation above are kept constant. This means tamping in the context of espresso preparation (where the pressure drop Δp is large) should be relevant to reducing fines migration.
- A shallower coffee bed depth L will be more susceptible to fines migration, with everything else in the equation above kept constant, simply because the drip rate will become much faster.
- The viscosity of water will affect local flow, but it is also involved in the equation that connects the local flow to the drag forces to which fines are subjected (e.g., Yang 2016, equation A-1). Because of this, water viscosity should have a limited impact on fines migration.

In general, a good approximation to keep in mind is that any suction or pump pressure will make fines migration much more of an issue in coffee brewing. This can result in cloudier brews or, in some situations, filter clogging. This seems to be true in the context of AeroPress brewing, for example: the harder one presses on the piston, the cloudier the resulting beverage seems to get. This has a noticeable impact on taste—although not necessarily a negative one, depending on your preferences.

The equation for the localized flow u we just discussed is, however, most useful to our intuition for scenarios in which we keep all the listed variables fixed except for one. For example, it does not make clear what happens if we tamp to reduce porosity but increase grind size to compensate and thereby achieve a beverage with the same average extraction yield. Doing so requires combining it with other equations in what we call a change of independent variables, which we will not study here. Koekemoer and Luckos (2015) noted that wider particle size distributions will produce slightly more packed beds, which may mitigate the effects of fines migration despite the larger amount of coffee fines.

An allongé shot pulled with the Decent espresso machine with coarse coffee grounds at very low pressure and a constant 1 g/s flow rate. An AeroPress filter was placed at the bottom of the espresso basket to filter out fines and oils, and a metal mesh was placed below the paper to ensure even flow through the paper filter. The resulting brew is far cloudier than a typical gravity-driven manual pourover because even that low amount of pressure is much stronger than gravity and caused even the smallest fines to migrate through the paper filter.

As an illustration of the concepts described above, it is useful to consider some allongés that Matt Perger brewed with tamped, nutated,[29] coffee ground midway between espresso and pourover grind sizes and a bottom AeroPress paper filter, using the Decent espresso machine in flow mode (constantly adjusting pressure to get a fixed drip rate of 1 g/s), corresponding to a pressure drop below 1 bar during the full brew. The goal of this experiment was to emulate the conditions of a manual pourover. He placed a metal mesh below the paper filter to ensure that water passed through the full filter surface rather than just the parts located on top of the basket holes. The resulting brews were much cloudier than anything brewed manually, which should not be surprising if we remember that gravity-driven brews are effectively using a pressure drop below 0.008 bars, a value so small that no espresso machine pump can maintain it reliably.

4.5 Evenness of Flow

How evenly water flows through the coffee bed is important because it will dictate how evenly the coffee particles are extracted. Obtaining an even flow is gen-

[29] a rotation and back-and-forth motion of the tamper during tamping that compresses the puck more than regular tamping

The flow of water is more spatially uniform through beds of even coffee particles (left) compared with uneven ones (right). Gradually increasing water velocities are represented with blue, green, yellow, and red colors. This illustration is based on a simulation by Ahmadi and Sefidvash (2018).

erally difficult in the context of percolation, for several reasons. First, uneven particle sizes will favor preferential paths of least resistance in the vicinity of coarse coffee particles, where water can flow downward in a more straightforward, less tortuous manner (Ahmadi and Sefidvash 2018; Holzbecher and Oehlmann 2012).

In addition to this, computer simulations involving loose and uneven porous media have demonstrated that fluids tend to flow more evenly at faster flow rates, an effect that is surprisingly encountered much before the flow is fast enough to become turbulent (Stanley et al. 2003). Adapting the data of these

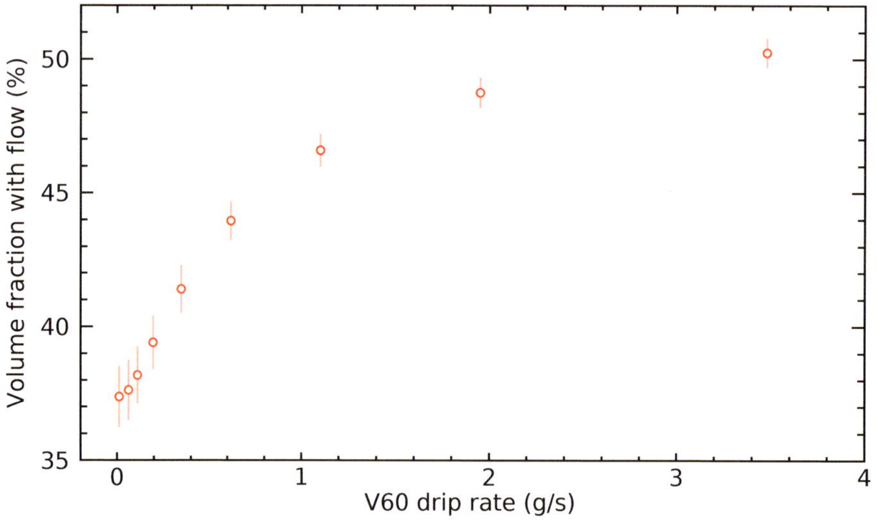

Fraction of the coffee bed that receives a non-negligible flow of water, as a function of V60 drip rate. This graph is based on data from Stanley et al. (2003) adapted to a typical V60 configuration.

authors to the parameters of a V60 brew allows us to see the fraction of the coffee bed that receives a non-negligible flow of water in the figure opposite, below.

These effects can also be compounded by the clogging of a paper filter or of the coffee bed itself following the migration of fines. This not only reduces flow, but it also leads to water taking preferential paths of least resistance and thus dramatically reduces the evenness of flow (e.g., see Dapp and Müser 2016). Compounding the problem, the flow of water that becomes too fast locally can displace the particles in a loose bed of coffee, creating a small region with larger pores and less hydraulic resistance. This can cause a runaway effect, where more water flows through and displaces yet more coffee particles, creating large channels. Rosti et al. (2020) built a limited model of such channels caused by elastic deformation of pores, by assuming particles can be deformed but not displaced. In practice, channeling tends to be more dramatic when we brew coffee that is ground finer, simply because the coffee bed has a lower hydraulic permeability. This increased resistance to the flow of water will more easily force water to flow along any channels forming in the coffee bed.

This effect could potentially explain why extraction yield does not seem to go up linearly as we grind finer, as would be expected based on a simple model

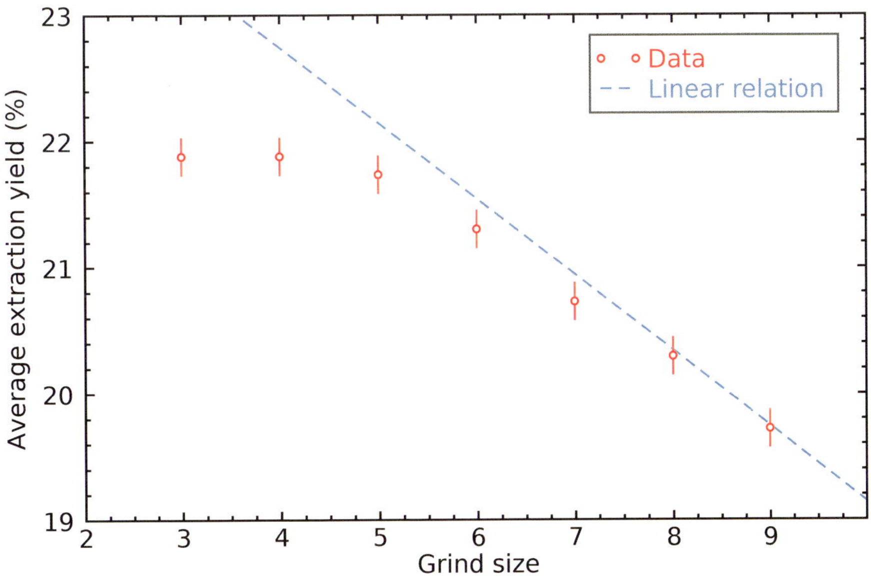

Average extraction yield as a function of grind size for the Ethiopian coffee Worka Sakaro roasted by Heart and brewed as V60, with the brew method described in Chapter 11. I typically use a grind size of 7.0, located close to the point where the observed average extraction yield departs from a linear prediction anchored on the coarser grind sizes.

with perfectly even flow through an immovable coffee bed that is free of channels or fines migration (Cameron et al. 2020). I observed this myself by brewing seven V60s in a row, changing the grind size by spacing the burrs ten 5-micron steps coarser with every subsequent brew.

Similarly, the stream of water falling on top of the coffee slurry can dig a hollow at the surface of a coffee bed and form a wide channel through which water will preferentially flow. This effect can be much more pronounced when we brew with a shallow coffee bed because any small perturbation of its surface will cause a more dramatic increase in hydraulic permeability, allowing water to flow easily and form a large channel. Using a coffee bed at least 3 centimeters in depth is good practice to avoid this (Rao 2008). Both an uneven coffee bed and localized channels can be mitigated by gently swirling the dripper after each pour. Doing so will tend to flatten the coffee bed and mix around the coffee particles to break any local channels that began forming.

While its effect is not as dramatic, a slanted work surface can also cause water to preferentially flow across one part of the coffee bed; I therefore recommend you verify that the counter on which you brew is relatively level, especially with brew methods that rely on gravity to drive the flow of water. Further problems of unevenness in flow can be caused by an improper preinfusion, as discussed above, or specific dripper design. Be aware that water bypassing the coffee bed or passing partially through its top layers can be the dominant

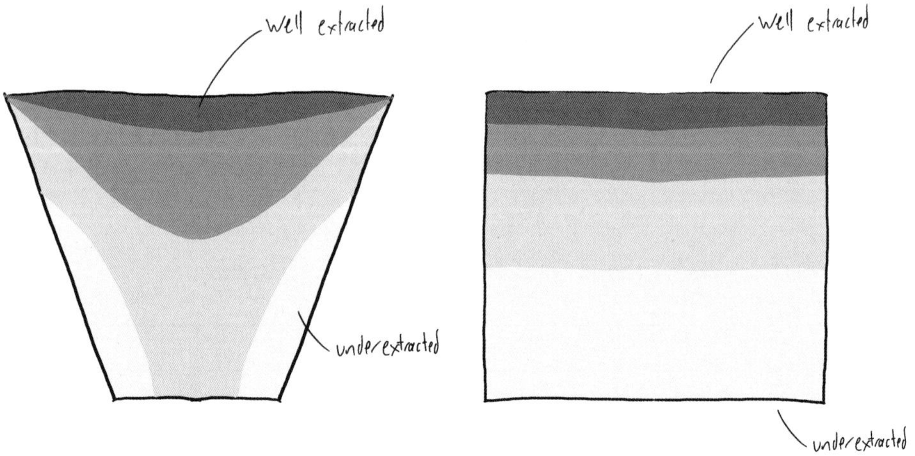

Spatial extraction maps for conical and cylindrical drippers at the end of a brew, assuming no water bypassed around the coffee bed or filter. The conical dripper has less horizontal evenness of extraction but more overall evenness because the bottom layers are slightly better extracted. In practice, however, this advantage is often reduced by water bypass or the clogging of paper filters. This drawing is based on simulations by Moroney et al. (2019).

cause of uneven extraction in some scenarios; we will discuss this in Chapter 7. Moroney et al. (2019) demonstrated that, in an idealized scenario with no agitation and no water bypass around the coffee bed, cylindrical drippers produce an extraction that is more even across the surface of the coffee bed because the velocity of water remains constant across it.

This was not true of conical drippers, but they found that conical drippers do produce extractions more even across the depth of the coffee bed. This latter fact is less intuitive; the effect is caused by the coffee being more concentrated, and therefore less potent at extraction, as it reaches the bottom layers of a cylindrical coffee bed. In the case of a conical dripper, more water will flow faster near the bottom of the coffee bed, partly countering the effect of it being a less efficient solvent. Even if this was engineered to obtain a perfectly even average extraction at all layers of the coffee bed, the exact flavor profile extracted from each layer would still be different, because using more water that is more concentrated will still favor the extraction of a different profile of chemicals. Agitation remains a very efficient way to make the extraction spatially more even in both cases.

4.6 Tapping and Vibrations

Vibrations can have a surprising effect on the structure of a loose, porous medium. Knight, Jaeger, and Nagel (1993) demonstrated that vibrations caused by repeated tapping in a container of dry spheres can introduce friction that carries particles around along cyclical currents. While this effect may be damped by the viscosity of water, any phenomenon that displaces the structure of coffee grounds in the slurry has the potential to favor fines migration, or even the formation of channels. The effect of tapping on the direction of cyclical flows of particles depends on the angles of the dripper walls and, potentially, on the frequency of vibrations, particle shapes, and frictional properties of the materials involved.

As these schematics on the following page show, the flow of particles introduced by the action of tapping forms a thin stream along the walls of the dripper; in the absence of water, particles larger than the thickness of this stream will not be dragged along and will get stuck at either the surface or the bottom of the coffee bed, depending on the direction of the cyclical streams. However, the presence of water introduces additional forces such as buoyancy, which can affect what particle sizes are preferentially carried to the surface. This often causes coffee chaff to accumulate at the surface of the slurry toward the end of a well-agitated brew.

This concept could be applied to purposely cause segregation by particle sizes in a dry coffee bed to systematically remove coarser particles. Much like the process of sifting, however, it would likely be difficult to repeat in a consistent and systematic way.

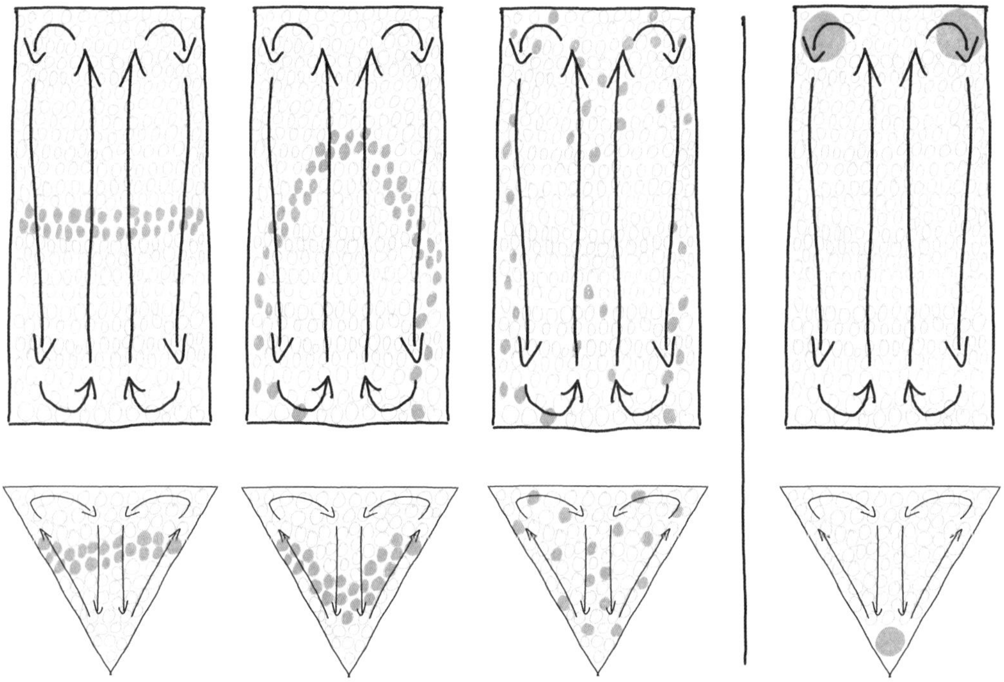

Schematics of cyclical flows induced in dry spherical particles by repeated tapping of cylindrical and conical containers. In the left section of the figure, a layer of colored balls is tracked over time, from left to right. The direction of flows in a cylindrical container is opposite to that in a conical container. Far right: A ball much larger than average will remain trapped at its contact point with the container wall because it is too wide to fit in the thin stream along the wall. Depending on the direction of flow, this phenomenon will cause larger particles to remain trapped at the upper surface of cylindrical containers and at the bottom of conical containers.

CHAPTER 5

Filters

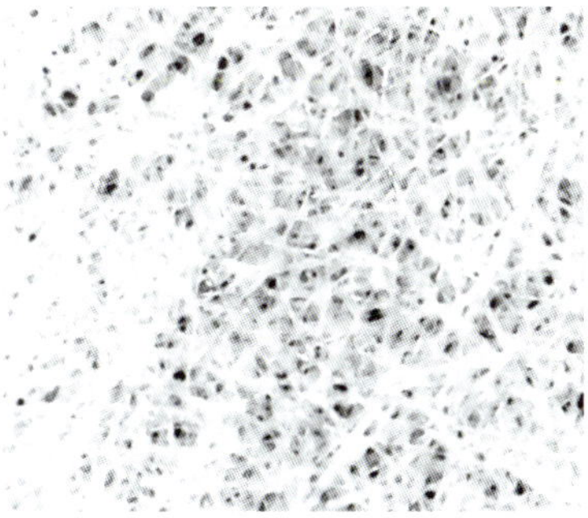

Filters play an important role in coffee brewing. They are responsible for keeping the coffee particles from reaching the beverage, while allowing a fast and uniform flow of water through the coffee bed. In this chapter, we review different types of coffee filters and their specific properties, as well as how they affect percolation and the resulting cup of coffee.

5.1 The Coffee Bed as a Self-Filter

Before we explore the properties of coffee filters, it is important to remember that for gravity-driven brews, the bed of coffee is our best filter to retain fines and produce a clean cup. We previously explored the concept of fines capture in Chapter 4, but a small experiment with the AeroPress brewer can demonstrate this situation even more clearly.

You can conduct this experiment yourself; prepare one AeroPress with a very fine grind, using the classic AeroPress recipe, and let the coffee cool down while you repeat the process, this time using a modified method. For this second brew, take the plunger off when you would normally push it through, and gently pour the coffee into a mug, from the upper side of the brewer. Try to get as little of the coffee grounds as possible in the mug. Be careful not to burn yourself, as the AeroPress slurry can remain quite hot. Clean the AeroPress, replace its paper filter with a new one, and then pour the contents of your mug back into the Aeropress chamber, stir the slurry, and *gently* press it through. Let the beverage cool for 15 minutes, and then compare the two brews, ideally in similar-sized glass containers.

Resulting brews from the AeroPress filtration experiment, with the same coffee. The brew on the left was poured out from the AeroPress after steeping and then post-filtered in a clean paper filter. The brew on the right was filtered normally through both the filter and coffee bed. The lack of filtration by a coffee bed made the brew on the left much cloudier. Photo by Andreas Corcaci

In the photo above, you can see the results my friend Andreas Corcaci obtained from this experiment: the second brew (on the left in the photo) is a lot cloudier because of fines and *colloids* (undissolved solids in suspension), which also give it a slightly more orange color. This can be explained by the absence of a percolation phase; only the filter was used to clarify the slurry, rather than a paper filter *and* a coffee bed. This is also why filtering immersion brews through a V60 or even a thicker Chemex filter barely improves their clarity. As we saw back in Chapter 4, using very gentle pressure (ideally, only gravity) is a good strategy to maximize the clarity of a coffee brew. The choice of filter will also have an effect, albeit a smaller one, on the clarity of a brew. If you carry out this experiment, be sure you press as gently as you can on the AeroPress; otherwise, both brews can become cloudy and wash out the effect of the coffee bed that this experiment aims to demonstrate.

5.2 Flow Rate

The rate at which water flows through a bed of coffee is in part controlled by the properties of the filter. Filters that are thinner or have larger pores will allow water to pass through faster. You may not have realized this, but coffee filters, with their maze of tiny spacings between paper fibers, are yet another example of a porous medium, much like a coffee bed. This is happy news, because it means we can immediately apply most concepts that we reviewed in Chapter 4 to coffee filters. For example, we can use Darcy's law to understand what is happening. If we consider a gravity-driven flow that goes only through a filter without any coffee, the drip rate will be

$$Q = \frac{k_\mathrm{f} A_\mathrm{f} g h}{\nu L_\mathrm{f}}$$

where the filter has its own values of hydraulic conductivity k_f, thickness L_f and area A_f. The use of different cross-sectional areas for the filter (A_f in the equation above) and the coffee bed (A in the regular Darcy's law) could seem surprising at first, but the surface of the paper filter through which water passes can be larger or smaller than the cross-sectional area of the coffee bed, depending on the filter and dripper designs. We will delve more into this last part in Chapter 7, when we discuss specific drippers. For a general reminder of other terms in Darcy's law, please refer to Chapter 4.

At this point, an informed reader may begin to wonder how that form of Darcy's law for filters interacts with the one we already studied, in the context of a lone coffee bed. This can be determined from fundamental laws of physics by noting that the total pressure drop across all percolation layers (the coffee and the filter) is equal to the sum of individual pressure drops and that the drip rate Q must be the same in all layers because water cannot be compressed at the pressures relevant to coffee preparation (e.g., see Kantzas, Bryan, and Taheri 2016). The details of this calculation, which are given in the Appendix, result in the following combined Darcy-like equation:

$$Q = \frac{g h}{\nu} \left(\frac{L_\mathrm{f}}{k_\mathrm{f} A_\mathrm{f}} + \frac{L_\mathrm{c}}{k_\mathrm{c} A_\mathrm{c}} \right)^{-1}$$

Here, the hydraulic conductivity k_c, the thickness L_c, and the cross-sectional area A_c are those of the coffee bed. Coffee filters are much thinner than typical coffee beds, but they usually have smaller pores. The difference in thickness dominates, however, so the combined effect is that filters typically offer less resistance to flow than the coffee bed does. Consequently, the filter properties generally have a relatively small effect on the drip rate: they cause a slight increase in hydraulic resistance as long as the effective filter surface A_f is not dramatically reduced by the dripper design. I know that equations are not always the most colorful way to visualize a concept, so I made a graph of the resulting drip rate Q for different filter resistances, assuming values typical of my V60 brews for all other variables.

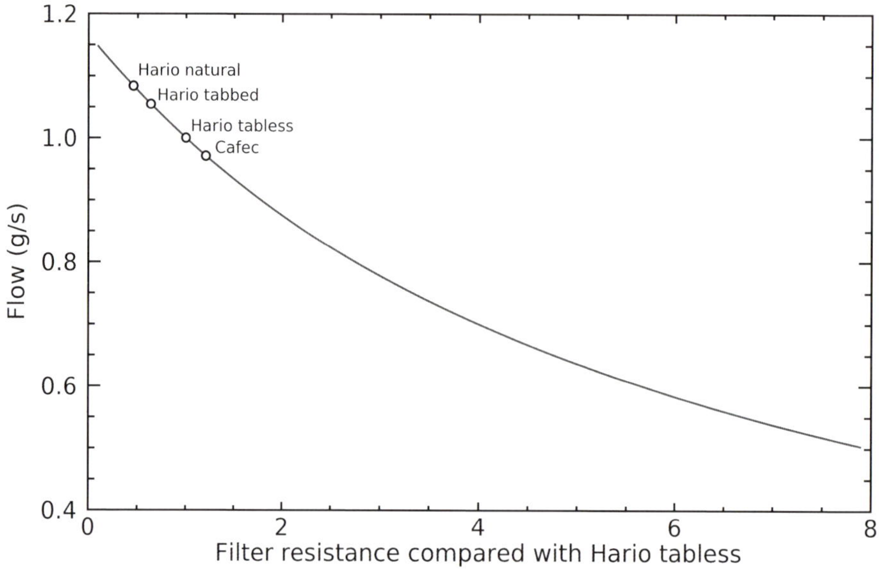

How the drip rate varies with the filter hydraulic resistance, as compared with the resistance of Hario tabless filters (set here at the reference point 1.0 in "filter resistance"). The horizontal axis has no units because it is relative to Hario tabless filters.

As you can see above, changing coffee filters has only a small effect on the drip rate because the changes in thickness and pore sizes are not dramatic from one filter model to the next. Reducing the filter area can, however, increase its resistance by a large fraction, which is why I extended the figure to large values of

A Hario tabless filter (left), and a Hario tabbed filter (right)

filter resistance. For example, the holes in the bottom part of an AeroPress cover about a third of the filter surface, which means that its hydraulic resistance is three times larger than what it would otherwise be if the filter was unobstructed. Keep in mind that this picture can change completely if a filter clogs. The figure assumes that no clogging takes place; we will discuss this topic later in this chapter.

The hydraulic permeability of a filter depends on both its pore sizes and how many pores it has per unit surface. This can be understood with the Hagen-Poiseuille equation, which specifies the discharge of water through a single cylindrical pore of cross-sectional area A and length L in the absence of turbulence:

$$Q = \frac{A^2}{8\pi\mu L}\left(\rho g h - \Delta p\right)$$

This equation is very similar to Darcy's law, but it depends on the square of the area instead of just the area. This is a consequence of the fact that water flows

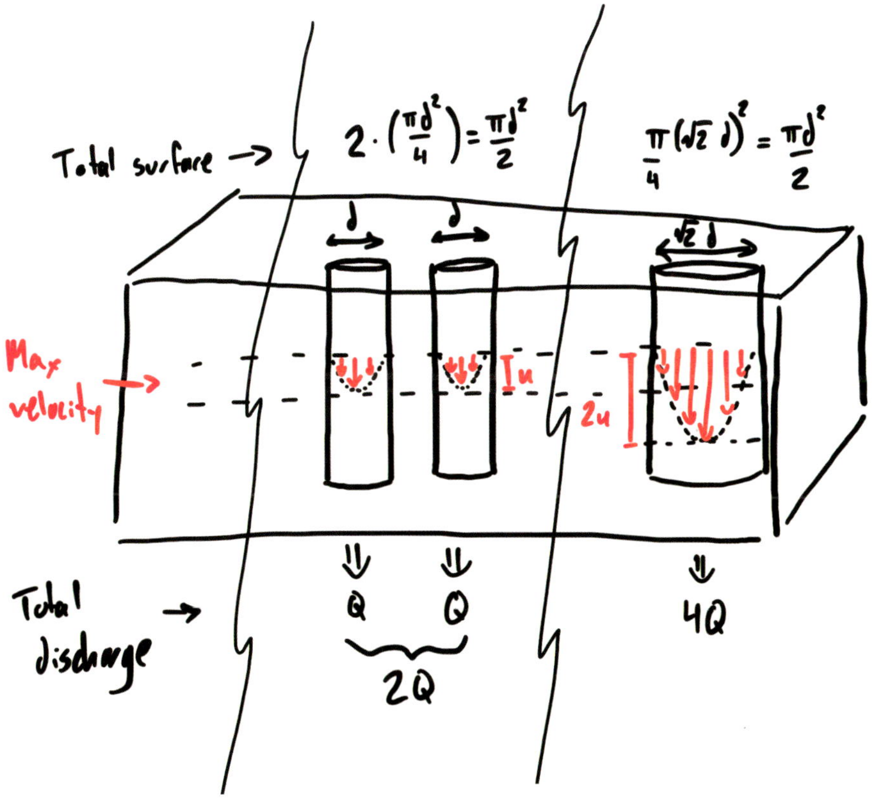

A schematic representation of the flow of water through two small pores versus one larger pore, both having the same total surface area. Because water flows faster (red arrows) on average in the larger pore, it will have twice as much discharge as the combined smaller pores, even if the total pore surface area is the same in both scenarios.

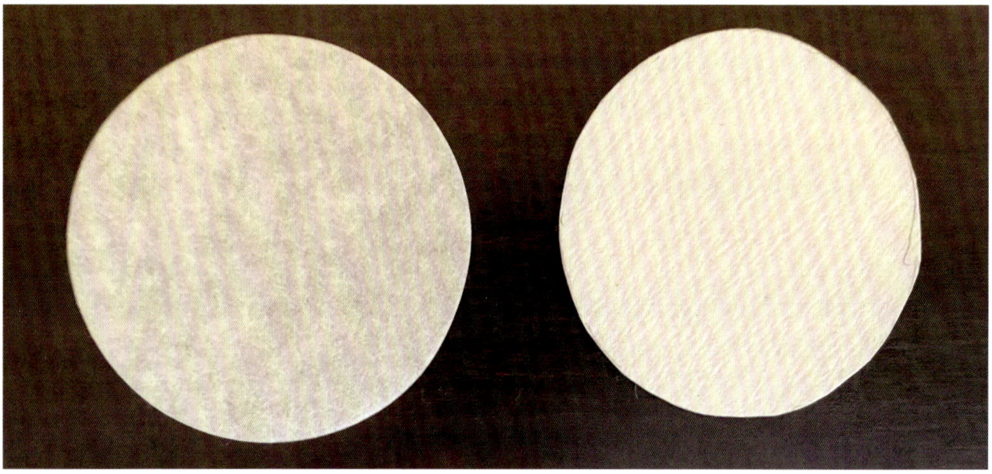

A smooth AeroPress filter (left) compared with a creped Whatman Grade 113 filter (right)

much faster at the center of a larger pore, compared with a smaller one. Therefore, having a single pore with twice as much surface area ($2A$) than two smaller pores (individual areas A, total area $2A$) will result in a faster drip rate.

This explains why doubling the surface of a porous medium only doubles its drip rate, as per Darcy's law. Doubling the surface area does not change the local velocity of water inside the pores; it only introduces twice as many pores where water can flow. In the simple case of a filter where all pores are circular and straight, with the same individual pore cross-sectional surface areas A_p over the full filter surface A_f, the Hagen-Poiseuille equation would allow us to determine the hydraulic permeability of the filter, as

$$k_f = \frac{\eta_f A_f A_p}{8\pi}$$

where η_f is the number of pores per unit surface area of the filter. In reality, most filter pores are noncircular and have tortuous shapes, which complicates the determination of their exact hydraulic permeability (Ghanbarian, Hunt, and Daigle 2016). The equation above still provides useful indications that having more pores per unit surface or larger pores will linearly increase the hydraulic permeability of the filter.

A nice trick can be used to increase how fast water can flow through a filter. Making the surface of the filter more irregular, through a process called *creping*, will increase the effective filter surface area of the filter A_f and therefore increase its hydraulic permeability. This is one reason why a lot of coffee filters on the market are creped.

5.3 Filtration

The main use of coffee filters is to keep the coffee bed contents together and separated from our cup of coffee. To do this, the filter only needs to prevent relatively coarse coffee particles from passing through; as discussed earlier, these particles will then take up the role of retaining even smaller particles. In some situations, the coffee bed fails to filter itself efficiently, and fines migrate down to the coffee filters. Two things can happen when such fines reach the filter: either the filter will let them through and produce a cloudier coffee, or it will stop them.

The ability of a filter to stop very fine particles depends significantly on the size and uniformity of its pores, but also on its thickness. As can be expected, filters with smaller and more uniform pores will be more efficient at blocking these small particles, but this comes at a cost because it will also reduce the filter's hydraulic conductivity and therefore the flow of water.

5.4 Load Capacity and Clogging

A coffee filter can stop fines that migrated down a coffee bed. This can be a good thing from the perspective of a barista who wants a clean beverage, but it can be the cause of another major problem: filter clogging. Effectively, every time a filter stops a particle from passing through, it is actually *storing* it somewhere among its maze of pores. It is useful to remember our discussion about fines migration in Chapter 4: after having dislodged and begun to migrate with the flow, fines can be recaptured either by coffee particles farther down the coffee bed or by paper fibers. Fines are usually stopped either by sticking to the side of a coffee particle or paper fiber with electrostatic forces, by getting lodged into a fold in their rough surfaces, or by completely blocking a pore smaller than the migrating fine.

The total volume of the region where water flows in a coffee filter gives us an indication of the amount of fines the filter can capture before it begins to clog, preventing the flow of water. This property is widely referred to as the load capacity of a filter. The total filtration volume, which determines the load capacity, can be expressed as

$$V_{\mathrm{f}} = A_{\mathrm{f}} L_{\mathrm{f}}$$

This immediately tells us almost everything we should know about the load capacity of a filter. Thicker filters will offer a better load capacity, and they will be more robust against clogging[30] (e.g., Sparks and Chase 2016), but this comes at the cost of their increased hydraulic resistance, as discussed previously, and can make the filters harder to prerinse properly. Increasing the effective surface area, for example by creping a filter surface, is another way to increase its load capacity. You might be tempted to think that drastically increasing the width of a cylindrical dripper is a good solution for increasing its load capacity, but this is not necessarily a good solution. In a wider dripper, the coffee dose will need to be larger to get the same coffee bed depth and avoid other percolation problems, which will also increase the total number of fines that the coffee can discharge into the filter. This is one reason why creped filters are such an ingenious feature: they increase the filter surface without changing the depth or the surface area of the coffee bed. Otherwise, the two most efficient ways to increase the load capacity of a filter are to maximize the effective filter surface by avoiding any obstruction of the filter and to increase the thickness of the filter.

A prepleated filter

Prepleated filters with folds on their sides are often used in laboratory equipment or in coffee preparation to maximize the filter surface area and give the filter walls more structure to prevent them from sagging. While this is great for filtering a solution in the context of chemistry, I do not believe it is a great option for the percolation of coffee. If water does pass through the folds of a prepleated filter, this means it did not pass through the entirety of the coffee bed, and this is not great for coffee extraction. Because of this, prepleating only increases the effective filter surface at locations where we actually don't want to be using it. This is not an issue when filtering a concentrated solution in a laboratory, because in this context the steps of extraction and filtration are usually kept separated. Prepleating also makes the filters, especially thick filters, dramatically heavier, and they will thus require more prerinsing to avoid a papery taste. They can also make it harder to disperse water evenly on the surface of the coffee bed, as some parts of the bed will be partly hidden through the folds, and pouring between the outside folds will guarantee that some water misses the coffee bed entirely.

[30] See also https://www.simada.co.il/images/filters/Whatman_filter_paper_guide_en.pdf

Clogging is not only an issue because the flow of water might stop before the desired brew ratio is reached. It is also a particularly bad issue because a filter will generally not clog everywhere at the same time, and this can cause a very uneven flow of water, even upstream and inside the coffee bed, which will lead to an uneven extraction (e.g., see Dapp and Müser 2016). This can make a brew harsh and astringent, to the point where it might completely mask an otherwise complex taste profile. I recommend interrupting a brew as soon as a filter begins to clog (where the drip rate falls much below 1 g/s) and adding more brew water directly in the coffee *server* to reach the desired concentration. This will typically result in an underextracted brew, which is usually not as bad as a harsh and astringent one.

Filter clogging can happen even in gravity-driven percolation, a context in which fines migration is usually less dramatic. It will more likely happen if a grinder or a particular type of coffee bean generates a larger amount of fines. One strategy to mitigate clogging is to maximize the ratio of effective filter surface to that of the coffee dose. The larger this ratio, the more filter pores are available to store coffee fines as they get plugged. This can generally be achieved by lowering the dose of coffee, as there will be less fines available to clog the filter. Other options that can help avoid clogging include grinding coarser; using creped and thicker paper filters with more load capacity; using a wide swirl during the bloom to store fines on the walls of the paper filter; not pouring water directly on the filter walls to avoid dislodging these fines; making sure the paper filter is not partially blocked by the dripper design; using a grinder that generates a more uniform particle size distribution; and sifting coffee fines. During his performance at the 2020 US Coffee Championships, Lance Hedrick of Onyx Coffee Roasters demonstrated that using a metal cone above a paper filter to lift the coffee bed mid-brew and then changing the paper filter is yet another viable option to maximize the effective filter surface.

5.5 Wet Strength

Another filter property, of smaller importance in coffee brewing, is the wet strength of a filter. This is an indication of how much pressure differential a wet filter can withstand before it tears apart. This applies mostly to paper filters, and even paper filters rarely tear when we brew coffee with gravity-driven flow. Wet strength can be more important when we place paper filters in an espresso basket or a Büchner funnel, but even then, the presence of a coffee bed above the filter generally limits the pressure drop that the filter itself is withstanding. For exam-

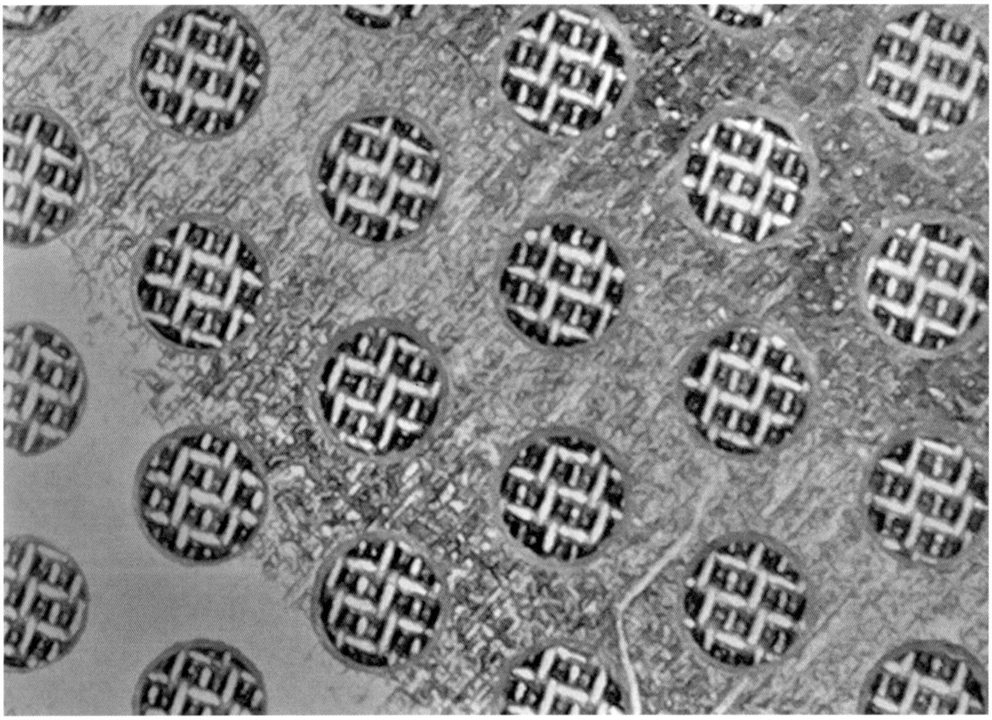

The Osaka metal filter for V60 viewed under a microscope. The image size is 43.2×28.8 mm with a resolution of 67.6 microns per pixel. All microscope filter images shown in this chapter have those same dimensions and resolution.

ple, the paper filter will not be alone in withstanding 9 bars of pressure drop in an espresso machine, because much of that pressure drop will take place across the coffee bed instead. Some approaches to increasing the wet strength of paper filters include making them thicker or reinforcing them with synthetic resins.

5.6 Material

The material that a coffee filter is constructed of is yet another aspect to which you should pay attention. Metal filters typically have enormous pores, and this will lead to a very fast flow and virtually no filtration of oil or coffee fines. Expect much faster brew times and beverages with more body and undissolved solids when you use these filters.

Paper filters will result in slower brews. They will absorb a fraction of the coffee oils as well as some of the fines that migrate out of the coffee bed, depending on the filter thickness. Most paper filters designed for coffee have relatively similar distributions of pore sizes; pore sizes are the spacings between the microscopic fibers of the paper.

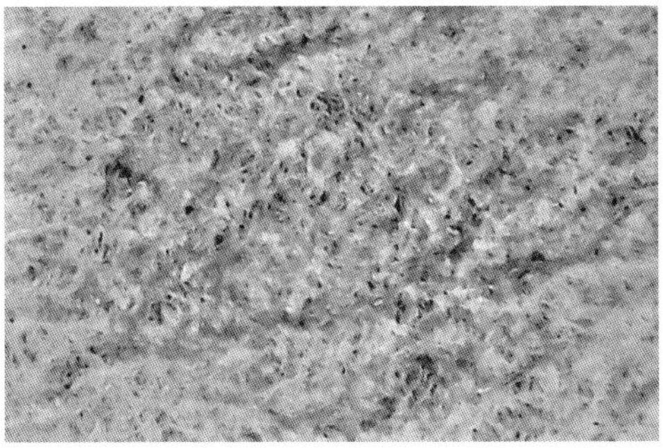
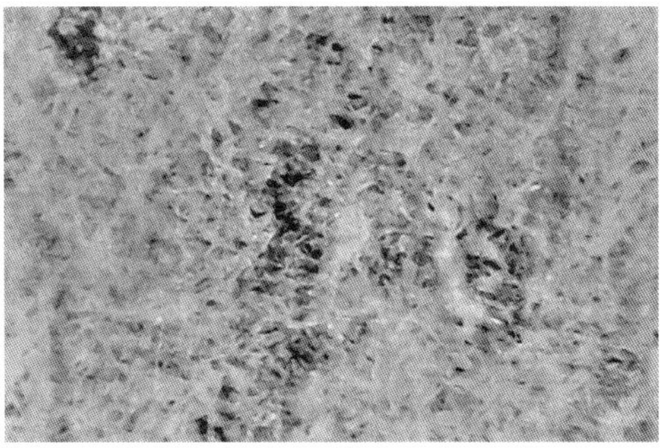

Hario V60 tabless paper filter version VCF-02-40W (left) and the more recent tabbed version VCF-02-100W (right). The average pore sizes are similar, although the original (tabless) filters have a better pore uniformity across the filter surface. The original, tabless filters are also creped on the inside (seen here).

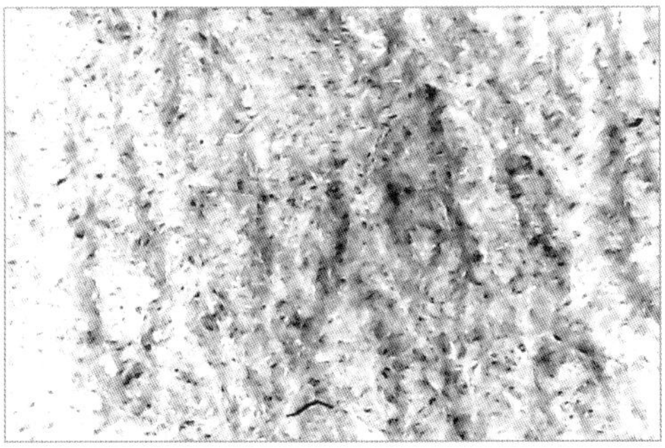
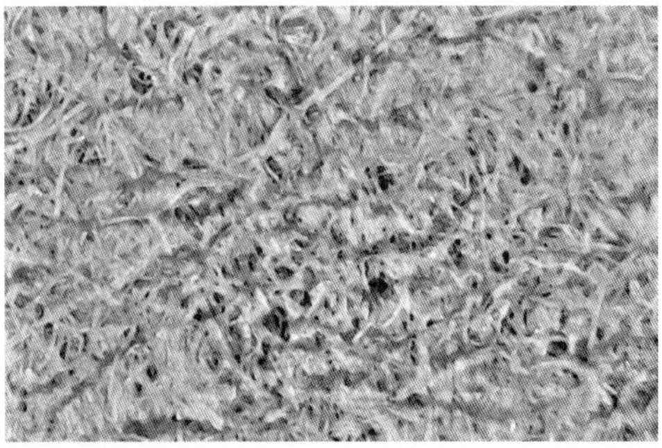

Hario V60 natural (unbleached) tabless paper filter VCF-02-40M (left) and Hario V60 natural tabbed paper filter VCF-02-100M (right). Unbleached filters retain more structure from the original paper fibers but can also impart a cardboard taste to the coffee unless they are rinsed thoroughly.

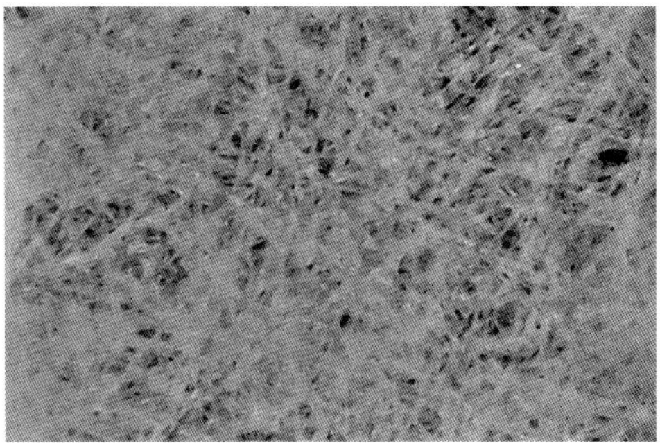

Chemex natural (unbleached) FSU-100 paper filter (left) and Chemex bleached FC-100 paper filter (right). Yet again, the fibers are better defined in the natural filters.

A microscope imaging analysis (described in detail in Chapter 11) resulted in the pore size distributions shown on page 117 (below), as a function of their relative contributions to the total filter surface.

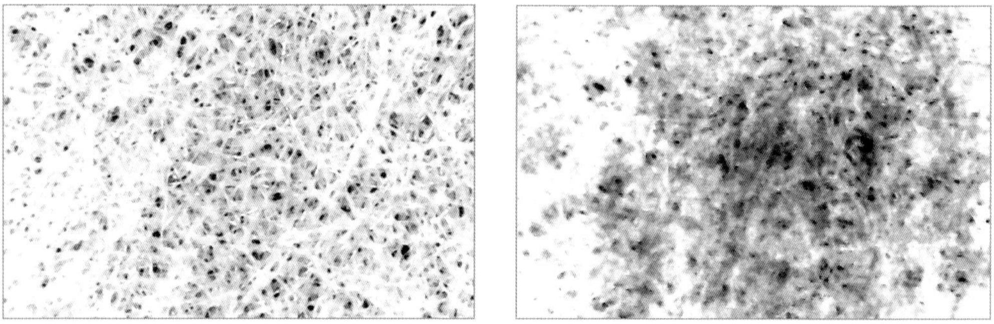

Sibarist paper filter (left) and Cafec universal AC4-100W V60 paper filter (right). The Sibarist filters have a larger number of pores compared with other pourover paper filters.

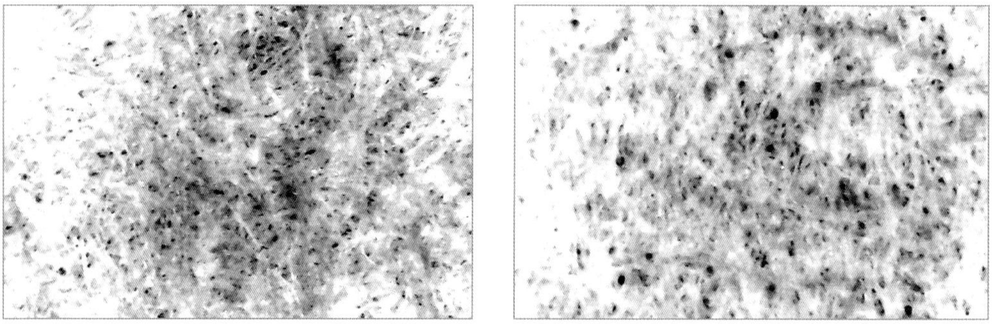

Cafec traditional paper filter (left), which come in a yellow package marked "Cafec white paper filters," and Cafec dark roast DC4-100 paper filter (right). According to Cafec, the dark roast filters are designed for an initially fast flow that slows down during the brew as the pores clog with coffee fines.

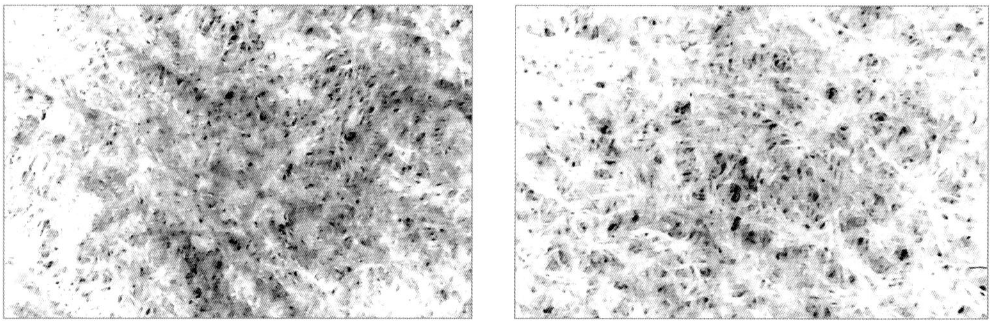

Cafec medium roast MC4-100 paper filter (left) and Cafec light roast LC4-100W paper filter (right). According to Cafec, the medium roast filters are designed for a sustained faster flow more robust against clogging, and the light roast filters are designed for a slower flow.

The figure opposite (below) shows that most pourover paper filters have similar pore sizes in terms of which ones contribute most to the total surface, although some have slightly wider distributions in pore sizes. It also shows that some filters have a larger number of pores.

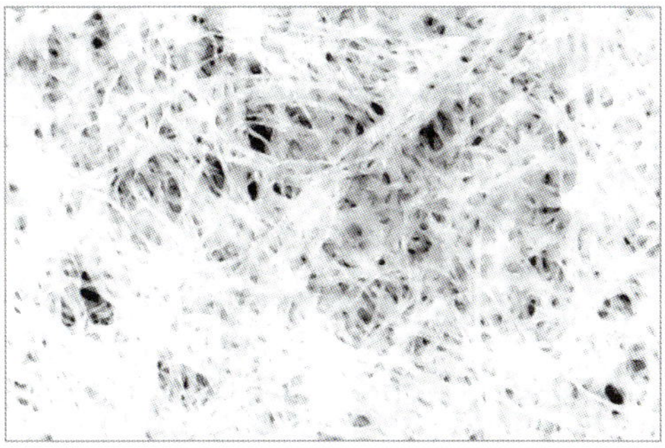

No brand (@_no_brand) paper filter (left) and Kalita paper filter (right). The No brand filters have dense fibers and opaque circular patches that seem free of pores.

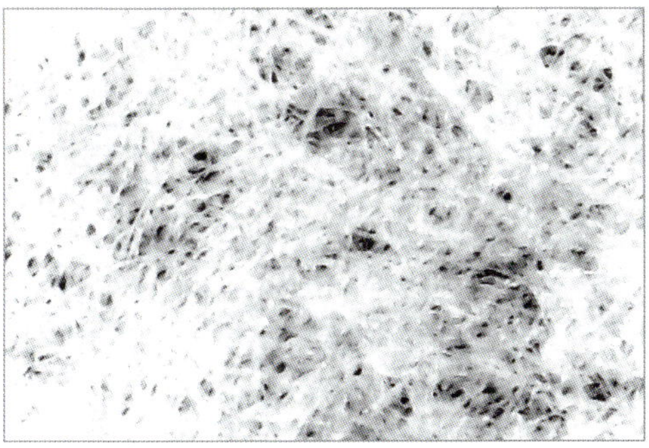

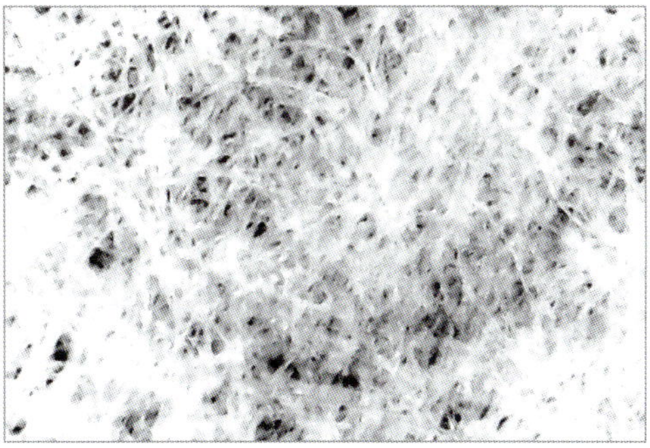

Fellow Stagg X paper filter (left) and Fellow Stagg XF paper filter (right) made for drippers with corresponding respective names

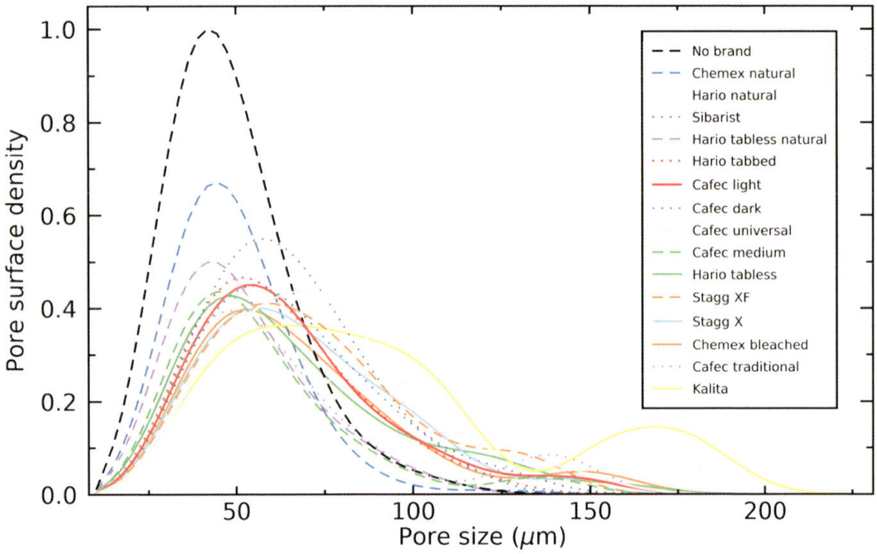

Total surface contribution per pore size for different pourover paper filters. While most distributions peak at about 45–50 microns, unbleached filters have narrower distributions of pore sizes and slightly smaller pores, on average.

Visualizing a normalized version of these distributions (below) will allow us to focus only on how narrow they are.

Bleached filters generally seem to have more uniform pores, but it is unclear whether this has a significant effect on the evenness of flow and on taste. It could

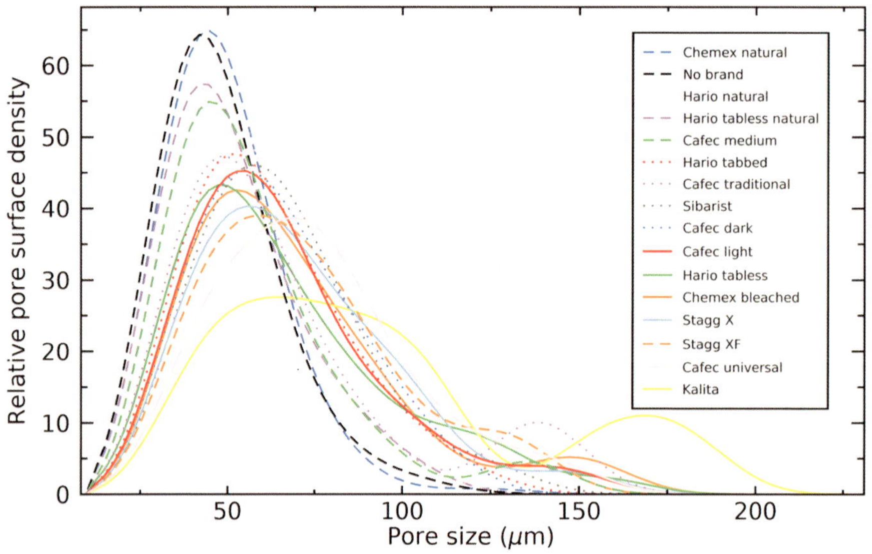

Relative surface contribution per pore size for different pourover paper filters. These normalized distributions allow us to focus on which filters have narrower pore distributions.

be expected that the resulting unevenness of flow propagates upward only a few millimeters in the coffee bed, because this is the typical scale of distance by which pores of different sizes are separated.

I measured the average drip rate of different paper filters without coffee grounds by placing a cut-out circle in the bottom part of a Delter press. I pre-rinsed them with room temperature (25°C, 77°F) water and then poured room-temperature water, using two scales to monitor the flow rate in real time. I inverted Darcy's law to calculate the hydraulic permeability of the filter divided by its thickness (a parameter I call β) and took the median and median absolute deviations during the full dripping event as the adopted values and measurement errors, respectively. The β parameters I obtained are shown in the graph opposite (above).

To translate these measurements into more easily interpreted units, I also produced a figure with the corresponding drip rates for a 1-cm-tall water column in the Delter press (opposite page, below).

Remember that these drip rates do not account for filter clogging. They are the combined effects of the pore size distribution and thickness of each filter. To illustrate this further, I repeated the experiment with 2 grams of decaffeinated

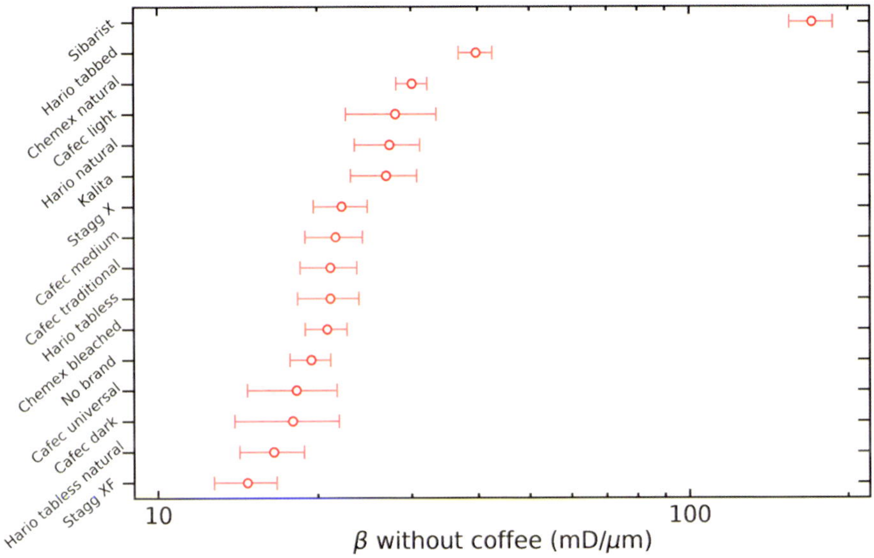

Hydraulic permeability divided by filter thickness (β) for all paper filters analyzed here, in units of millidarcy per micron. Larger values of β correspond to filters that flow faster for a given weight of water in the dripper and exposed filter surface.

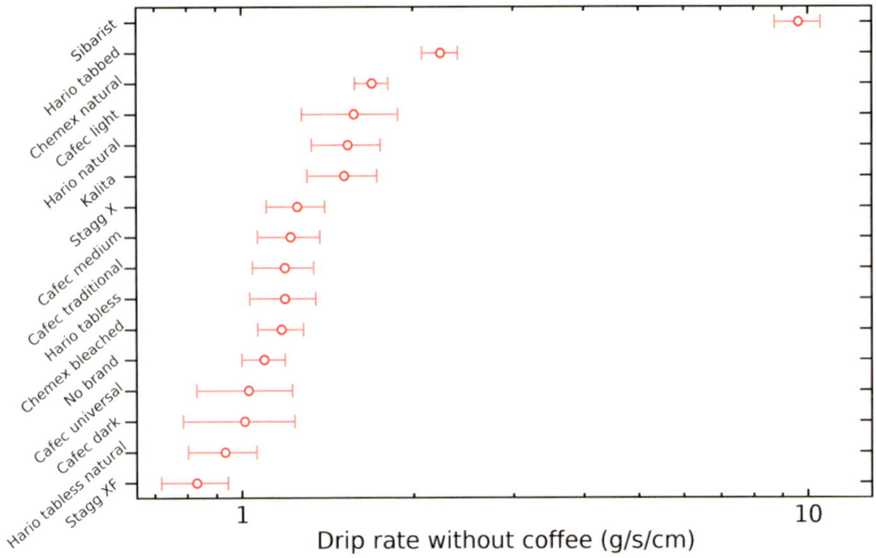

Drip rate per centimeter of water column height in the Delter press

coffee (Kenya Lenana AB Juicebox from Lüna Coffee) and a grind size typical of V60 brews placed in each filter before taking the measurements. The resulting β values and corresponding flow rates are drastically reduced, indicating that some pores were blocked by coffee fines.

These results are consistent with the manufacturer specifications by Cafec, as they specify that their medium roast paper filters are designed to sustain a higher flow for the full duration of the brew. It seems that they are indeed among the fastest paper filters in the presence of ground coffee.

Careful observers will have noticed that the order in terms of which filters flow slower or faster have changed after adding coffee in them. This is because not all filters clog by the same amount. To show this further, I defined a clogging

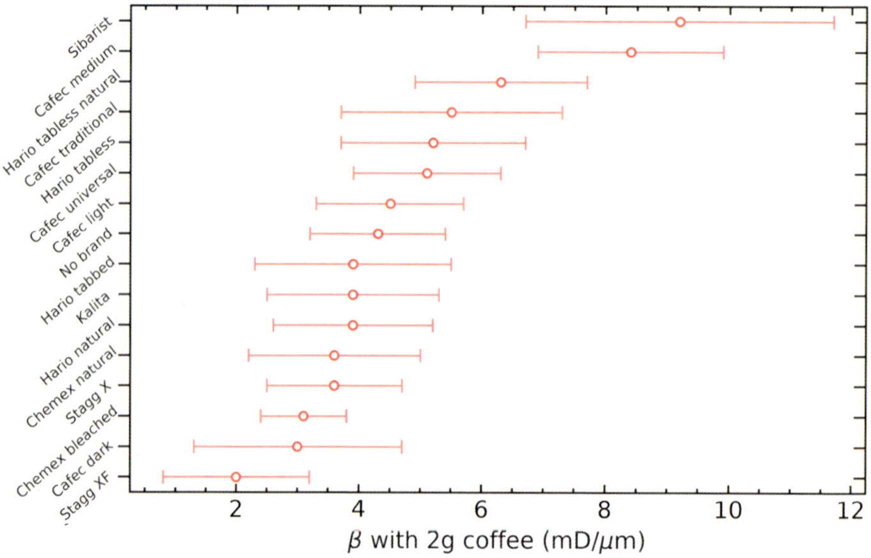

Hydraulic permeability divided by filter thickness (β) after adding 2 grams of decaffeinated coffee

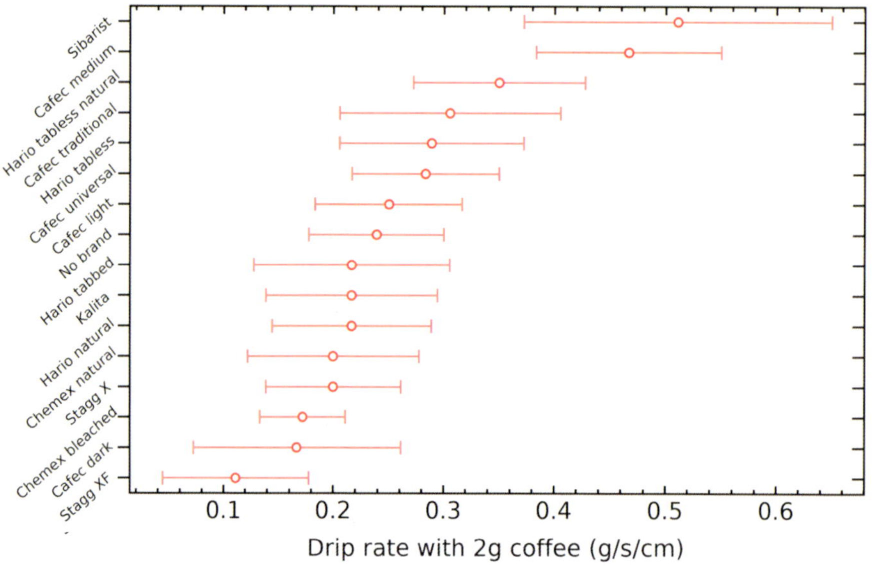

Drip rate per centimeter of water column height after adding 2 grams of decaffeinated coffee

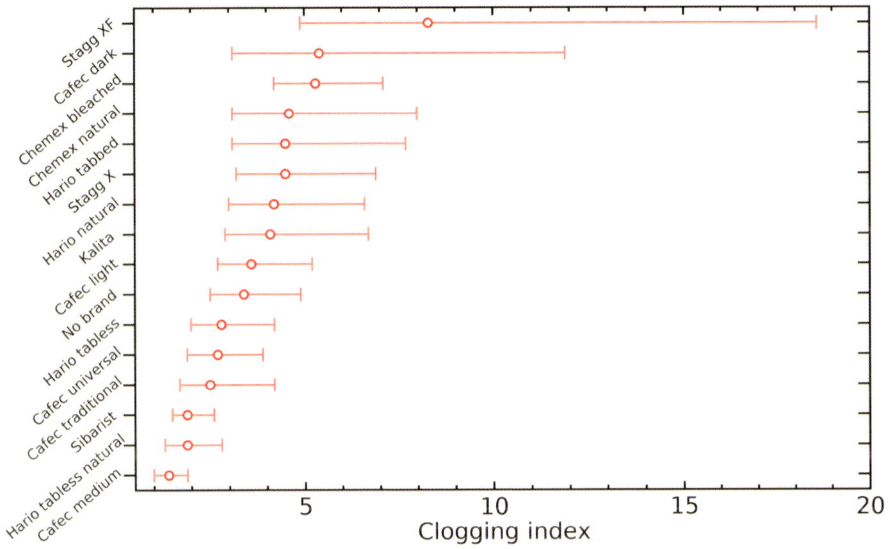

Clogging index for different pourover paper filters

index corresponding to the number of layers of Hario bleached tabless filters that would equal the added hydraulic resistance of the 2 grams of coffee.

As we discussed earlier, how prone filters are to clogging is related to both their pore distributions and their thickness. It is interesting to note that the Cafec dark roast filters are among the most prone to clogging, and this was specified as an intended property of these filters by the manufacturer to slow down the flow near the end of the brew.

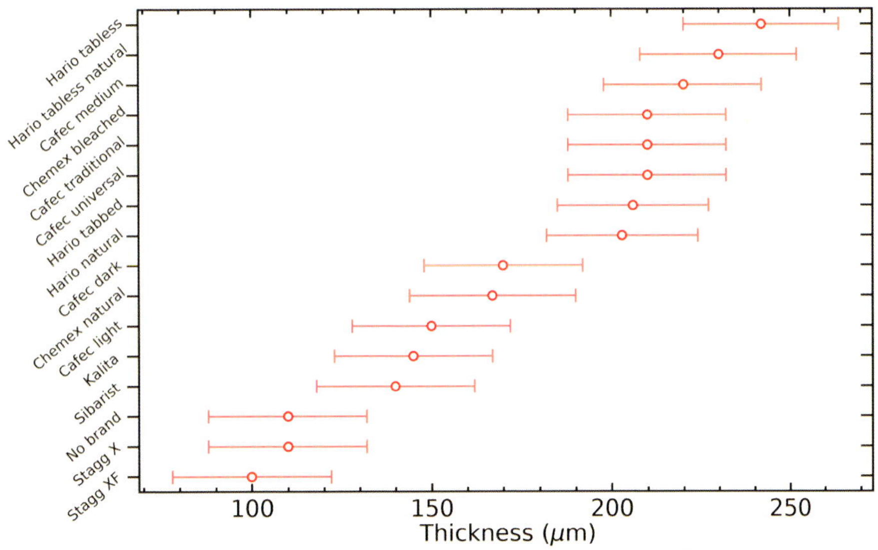

The thickness of different pourover paper filters

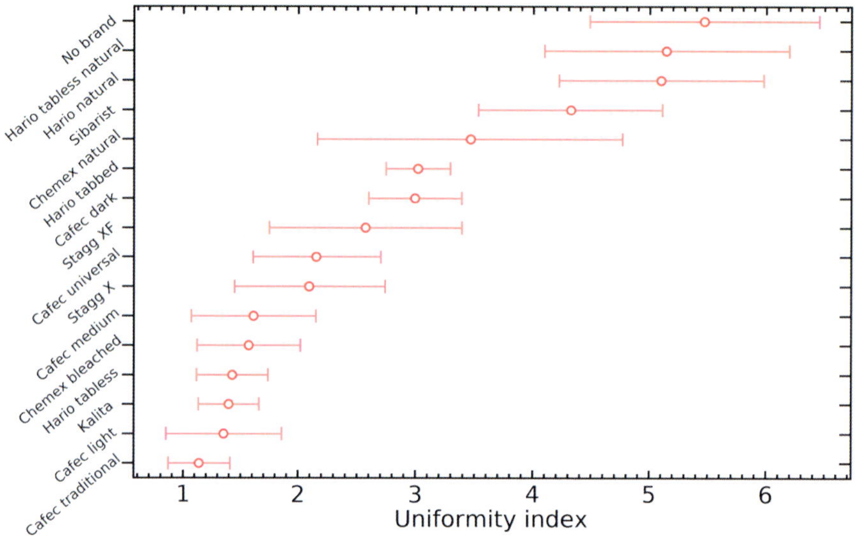

The uniformity index of different pourover paper filters, which indicates how evenly pores are distributed across their surfaces

I measured the thickness of these filters, as shown on the previous page, with a 20-micron precision digital caliper. Because clamping the caliper onto a paper filter can compress it, I clamped it gently and repeated each measurement at least thirty times to determine a reasonable average and measurement error.

It is important to note that some filters, such as the Stagg X and XF, may seem thick to the touch, but in reality they are among the thinnest paper filters available for pourover coffee. They are more rigid than other filters, giving us a false impression of thickness.

As a last characterization of the properties of paper filter pores, I used the thickness values above and my microscope images to calculate how fast a drip rate one would expect in each specific image using the Sampson-Poiseuille equation,[31] and I calculated how much this expected flow varies from one image to the next of the very same paper filter. The characteristic variation in expected flow divided by the average expected flow is thus a measure of how evenly the pores are distributed across the surface of the filter, a different consideration from how evenly sized the pores are. I took the inverse of this quantity as a uniformity index, such that larger values correspond to filters with pores more uniformly distributed across their surfaces. The results are shown above.

Bleaching is another important aspect of paper filters; either chlorine or

[31] This assumes the pores are circular and straight down, which is not the case but will provide a valuable first estimate.

oxygen is used to remove the brown color of "natural" paper filters. This drastically reduces the cardboard or papery taste that they will impart to water. I recommend using bleached filters for this reason, as well as prerinsing them with hot water to be sure you avoid such paper flavors. Some bleached filters will almost never produce cardboard flavors, but this is not true of all filters that look white. Thicker and larger filters will require more rinsing water, and unbleached filters will require yet significantly more. As Rao (2010) recommended, you can prerinse a paper filter and then immerse it in hot water for five minutes, let it cool, and then taste the water to determine whether your rinsing procedure is appropriate for that particular filter. I recommend doing this blindly with another cup of water that never touched the filter to ensure you are actually detecting a flavor.

As a demonstration of this effect, I rinsed different filters by placing 200.0 grams of distilled off-boil water on the dry filter in a Hario SWITCH immersion dripper[32], placed a cork lid on top for heat insulation, and waited three minutes before turning on the flow of the Hario SWITCH. I let the water cool down to 27°C (81°F) in a covered ceramic cup and then measured the approximate total dissolved solids of the water with an electric conductivity meter. Among the four filters I tested, I was able to detect a very small quantity of dissolved solids only in the Cafec universal filter and the Hario natural filter.

- Hario tabbed bleached VCF-02-100W: 0 ppm
- Hario tabless bleached VCF-02-40W: 0 ppm
- Cafec universal bleached AC4-100W: 1 ppm
- Hario natural VCF-02-100M: 5 ppm

Because I showed a lot of different measurements above, I decided to build two figures that aim to summarize most of them, shown on the following page. The first figure shows filter thickness as the horizontal axis, and the second shows the filter uniformity index instead.

The latter of the two figures is interesting because it shows that all of the natural filters have a systematically better distribution of pores across the filter surface as well as more uniform pore sizes (we showed this part earlier). It is unclear to me how important this effect is on the uniformity of flow across the full coffee bed, but using natural filters definitely risks imparting a cardboard taste to coffee.

Cloth filters are also available for coffee filtration, but they are, frankly, a nightmare to use. They result in fast and cloudy brews, similar to those the metal filters produce, but they also absorb some coffee oils. This can be pleasant at

[32] a device similar to a Hario V60, but that allows the user to turn flow on or off at will with a switch

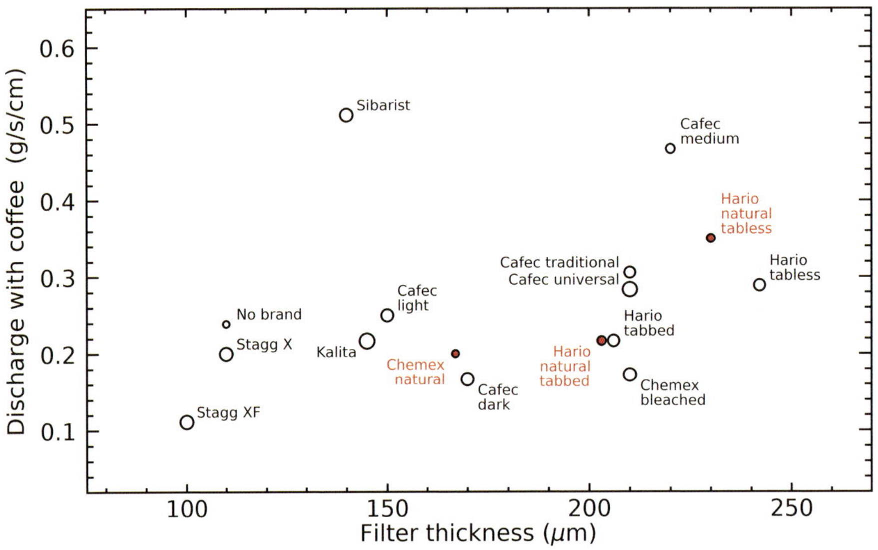

A summary of properties for all pourover paper filters discussed. The symbol size corresponds to the fraction of pores larger than 50 μm (a proxy for how much body the beverage will have if one also considers filter thickness), and the symbol color corresponds to whether a filter is bleached (white) or natural (dark brown). The Hario tabless bleached filters I use daily are thick and yet maintain a good drip rate in the presence of coffee fines.

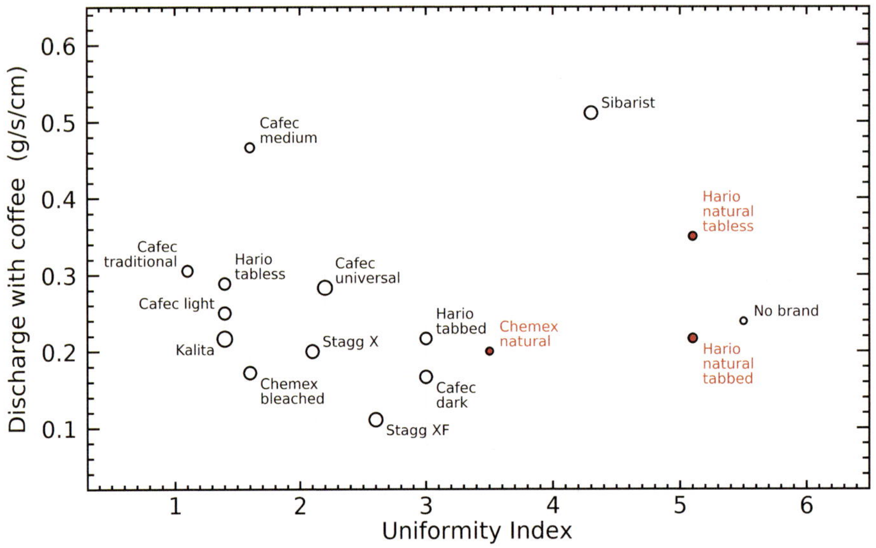

A different summary of properties for all pourover paper filters discussed. The symbol size corresponds to the fraction of pores larger than 50 μm, and the symbol color corresponds to whether a filter is bleached (white) or natural (dark brown). Natural filters have better uniformity of pores across their surfaces.

first, but it will rapidly cause the filters to impart a rancid taste to coffee brews, due to the stale oils that they accumulate. Things can get even worse if the filters are not rapidly cleaned and dried after use, as bacteria or fungi can colonize them and impart a taste that evokes the smell of old rags. This tastes as good as it sounds.

Siphon brewers often come with cloth filters, but adapters for paper filters are also often sold separately; I highly recommend using paper filters if you brew with a siphon. If you insist on using cloth filters, I recommend getting at least a dozen, washing them immediately after use, placing them in a bag in the freezer, and collectively boiling them in water with odorless detergent (such as Oxiclean) after each filter was used once. They will need to be boiled again two or more times with clean water to remove the taste of detergent and immediately hung to dry. As you have probably surmised, this process is impractical, especially if you live in a humid environment.

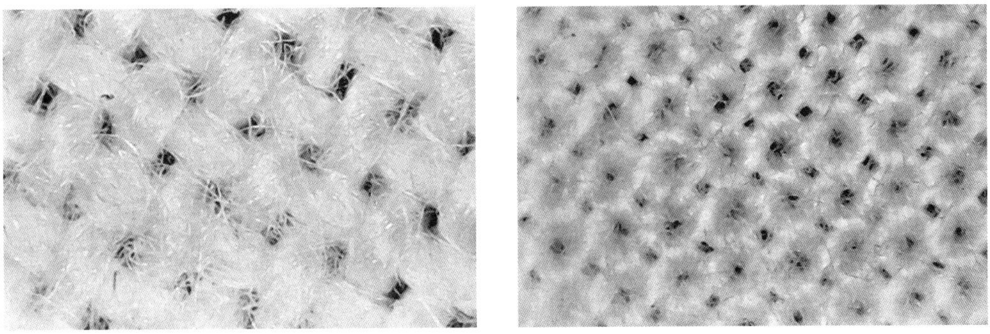

A Hario siphon cloth filter (left) and a V60-02 "CoffeeSock" cloth filter (right)

CHAPTER 6

Kettles and Agitation

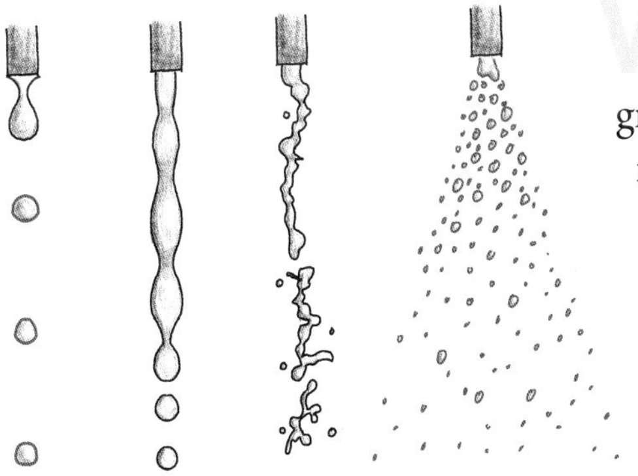

While a kettle does not impact coffee brewing anywhere as much as a grinder, choosing a better kettle can afford more control on how water is precisely dispersed onto the slurry. This can be important for achieving better consistency from one brew to the next, and it can allow more control of brew time and agitation. Other considerations, such as heating efficiency, water capacity, and noise, can be valid considerations to guide the choice of a kettle, but we will not consider those specifically; we focus in this discussion on the technical aspect of brewing. This chapter first considers some underlying physics to help us understand how a kettle affects brewing, a quest that will take us to surprising corners of the scientific literature. This will give us a deeper understanding that will allow us to discuss how to best select and use a kettle.

6.1 The Teapot Effect

Traditional kettles usually have a wide spout, making it difficult to pour water in a gentle and controlled way. Water tends to flow along the spout and the outer wall of the kettle rather than along a well-detached stream, worsening this effect. Adhesion forces between water molecules and the kettle material cause the water to flow along the kettle unless the water is poured at a fast-enough rate. The appropriately named teapot effect is illustrated on the following page.

There are different ways to mitigate this effect. A thin and elongated

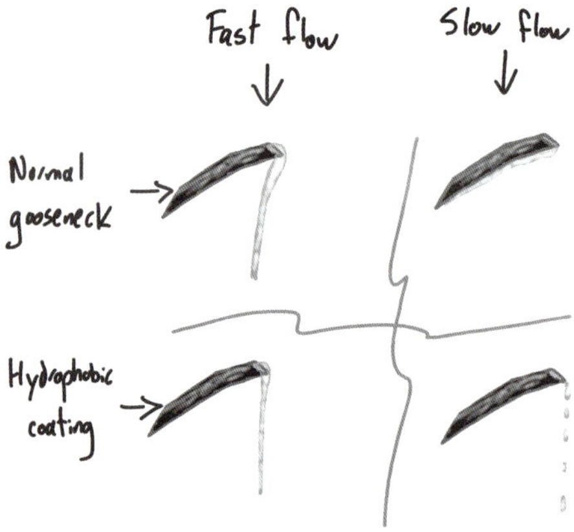

An illustration of the teapot effect (top row) and how it can be mitigated by a hydrophobic coating (bottom row)

Photo of a water droplet forming on the hydrophobic coating of a gooseneck kettle. If you remember the concepts of water cohesion and adhesion from Chapter 4, you may recall that removing adhesion between water and the kettle with a hydrophobic coating will cause the water to form droplets because water molecules cohere to (are attracted to) other neighbor water molecules.

gooseneck with an appropriately shaped spout can reduce it for all but the slowest pours, but by far the best way to eliminate the teapot effect is to coat the tip of the gooseneck with a hydrophobic material to completely remove adhesion forces (Duez et al. 2010). To my knowledge, only the Fellow pourover kettles currently use this technology, and they allow pouring at an extremely low drip rate, to the point where only a stream of tiny droplets is falling, without any sign of the teapot effect. In the photo shown at left, we observe a droplet of water forming at the tip of the gooseneck kettle, indicating that the outer surface of the gooseneck is hydrophobic.

6.2 Breakup Length

Even in the absence of the teapot effect, no kettle produces a perfectly coherent stream of water, especially when the user is pouring very slow or fast. Small imperfections at the tip of the gooseneck will cause the stream of water to have a slightly irregular shape at the top of the stream. The surface tension[33] of water will cause inward forces at the surface of the stream that vary depending on its local curvature, and those inward forces will cause these imperfections to grow gradually larger as the stream falls farther. This is an example of a hydrodynamic

[33] The surface tension of water, discussed in Chapter 4, is a consequence of cohesive forces between water molecules larger than adhesive forces between molecules of water and those of the surrounding air.

Illustration of Rayleigh-Taylor hydrodynamic instability, responsible for the shapes often seen when smoke is propagating through air (based on a simulation by van Heerwaarden 2014)

instability, a wide class of physical phenomena whereby small imperfections become significantly amplified, giving rise to a plethora of beautiful fluid motions, including cloud patterns at the surface of Jupiter, shocks of gas that follow supernovae explosions, and patterns formed by smoke (described by the Rayleigh-Taylor instability).

The gradually deforming stream of water is an example of Plateau-Rayleigh instability. Visually, this causes the stream to display wavy patterns at its surface that become gradually larger as the stream falls. When the waves become as wide as the stream itself, the flow of water separates into droplets.

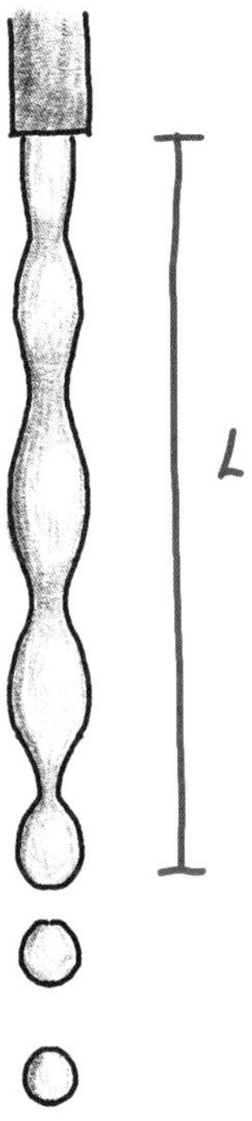

An illustration of the Plateau-Rayleigh instability, breaking up the water stream after a length L

The point at which the stream separates into droplets is called the breakup length[34] (e.g., see Nath Mishra 2018), an important concept for coffee brewing. This specific type of instability happens only for a given range of pour rates, depending on the thickness of the kettle's gooseneck. Within this specific range, *the breakup length is longer for faster pours rates*. Pouring very slowly will cause the water to come out as droplets, with an effective breakup length of zero; pouring extremely fast will cause new instabilities that are technically described as wind-induced, or even atomization in the most extreme cases. Both of these very fast flow regimes have

[34] This is also sometimes called the length of coherence.

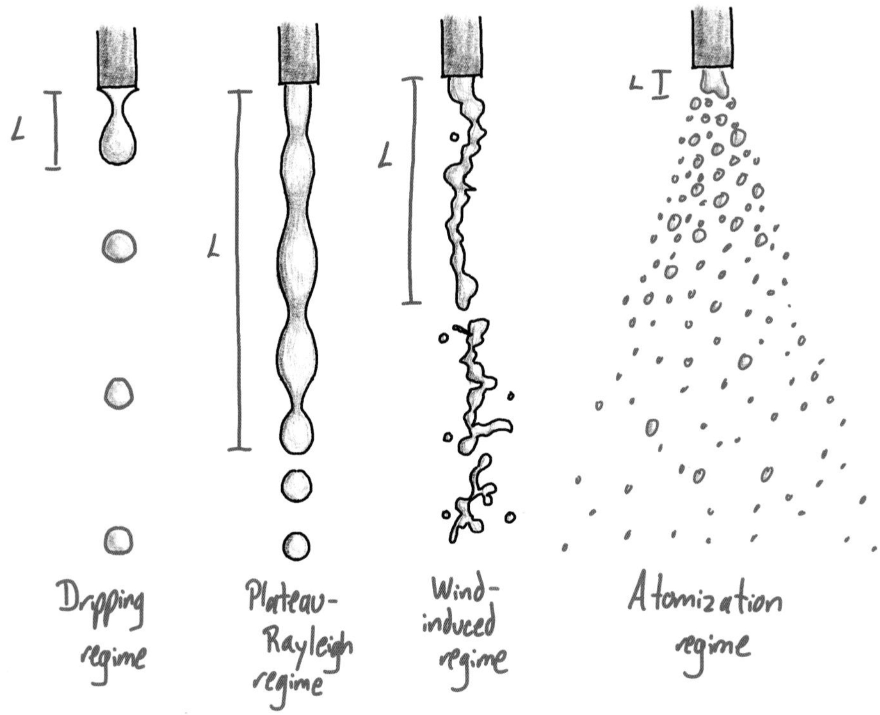

Illustration of different flow regimes in increasing order of flow velocity, with their respective breakup lengths *L*. This schematic was adapted from data by Bertola and Brenn (2019).

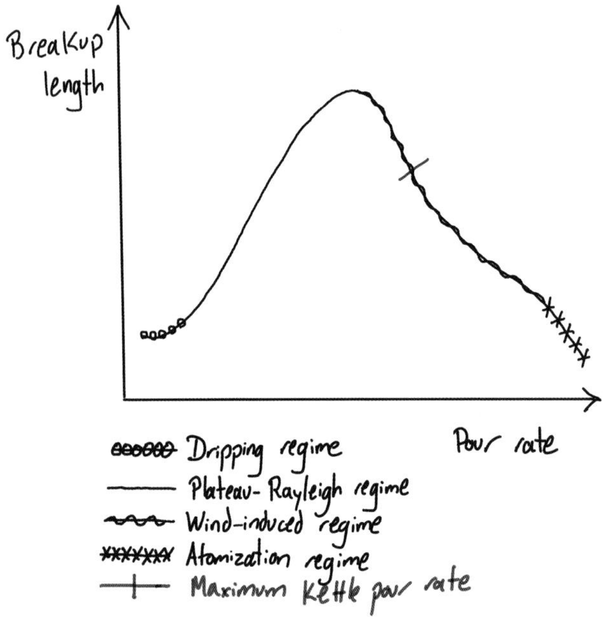

Breakup length as a function of pour rate with the respective flow regimes indicated. A perpendicular line is placed where typical pourover kettles reach their maximum pour rate.

shorter breakup lengths, as illustrated above.

I have never encountered a kettle that can reach the atomization regime because kettles' pour rates are limited by the weight of water inside the kettle. Rather, the maximum pour rates usually seem to be located somewhere after the transition to the wind-induced regime. An example curve describing how pour rate affects breakup length is shown at left.

Precision kettles such as the Brewista Artisan and Fellow Stagg EKG have thinner goosenecks and will therefore have their longest breakup at relatively slow pour rates. By comparison, the longest

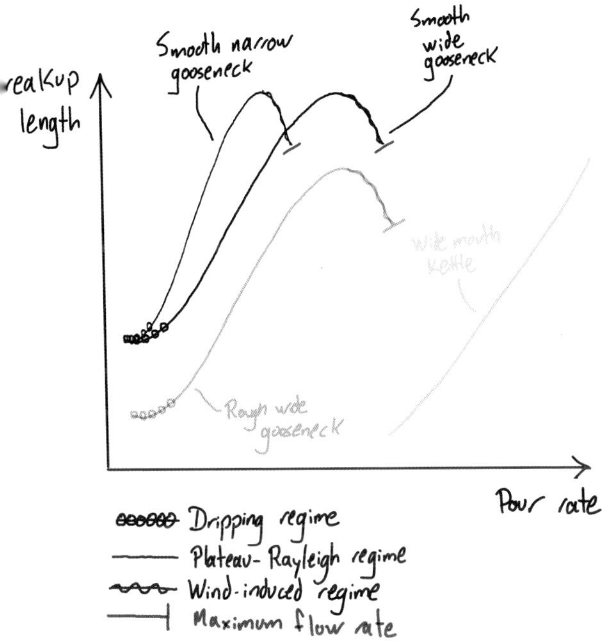

Effects of gooseneck smoothness and diameter on the breakup length relation, cut off at the maximum pour rates of the respective kettles

breakup of a wider kettle such as the Bonavita will happen at fast pour rates. In other words, changing the diameter of the gooseneck will horizontally stretch the breakup length versus pour rate curve. In addition to this, kettles with smoother gooseneck openings will have longer breakup lengths, on average. These effects have been well studied by engineers in the context of spray coating processes (e.g., Jadidi, Moghtadernejad, and Dolatabadi 2015). The diagram at left shows how gooseneck diameter and smoothness affect the breakup length versus pour rate curves.

I verified these relations experimentally with the Stagg EKG 0.6-liter and Brewista Artisan 0.6-liter gooseneck kettles, the latter with and without a flow restrictor that reduced the diameter of the gooseneck base from 5.5 mm down to 4.0 mm. I poured room-temperature

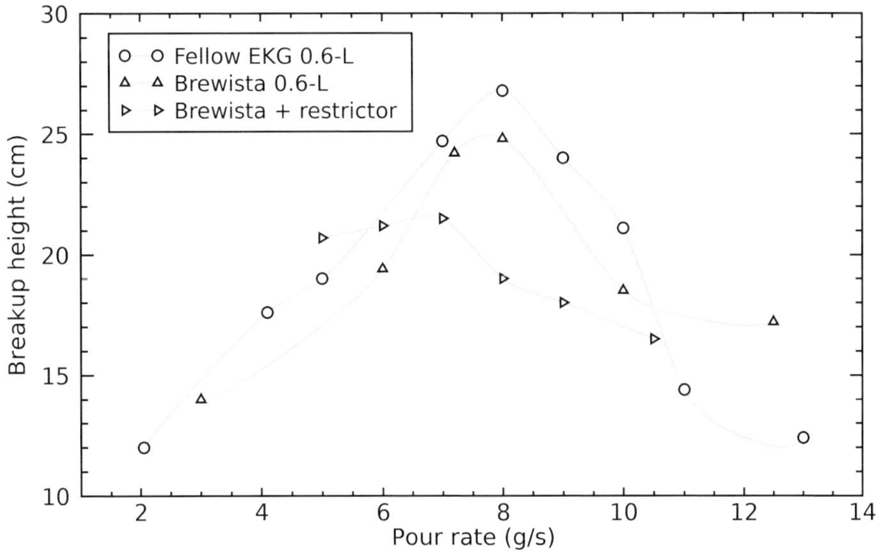

Pour rate versus breakup length with room-temperature water for the Fellow Stagg EKG 0.6-L and Brewista Artisan 0.6-L kettles, the latter with and without a flow restrictor

water at different rates, measuring live on an Acaia Pearl S scale, and then adjusted the pour height to determine the exact point when I began hearing a splatter sound. I then stopped pouring and measured the height of the gooseneck tip as well as that of the slurry, the difference of which I took as the breakup length.

The resulting data are very well matched by the physics literature across the dripping regime to the wind-induced regime. They indicate that the Stagg EKG has a slightly smoother stream, due either to the shape of the gooseneck tip or to its smoothness. The flow restrictor reduced the optimum pour rate of the Brewista 0.6-L kettle where its breakup length is maximal, but it also reduced its maximum breakup length. The full, usable pour ranges of these respective kettles were 1–18 g/s (Stagg EKG 0.6-liter), 2–35 g/s (Brewista Artisan 0.6-liter), and 2–15 g/s (Brewista Artisan 0.6-liter with a flow restrictor). Note that breakup lengths may shorten at higher water temperatures.

At this point, it would be fair for the reader to wonder why this is of any interest in coffee brewing. The reason is that the exact shape of the water stream will determine how much agitation is imparted to the coffee slurry and how deep that agitation reaches.

6.3 Agitation

To illustrate the effect of agitation in coffee brewing, it is useful to consider devices such as the Melodrip that allow a barista to remove virtually all agitation. This is achieved by splitting the kettle stream across many small openings, reducing the pour rate of each opening back to the dripping regime and thus effectively reducing the breakup length of any kettle to zero. The result of this removal of agitation will be a brew where the coffee bed remains intact and undisturbed but where the average extraction yield is significantly reduced, if no other parameters are changed.

These devices are normally used in combination with either many pours or much finer grind sizes to offset the loss of agitation, but the resulting brews will be very different from a coarser brew prepared with agitation of the grounds. A brew with more pours will tend to have a lower slurry temperature, and grinding finer will make it harder to achieve an even flow of water. Furthermore, as we saw in Chapter 1, using either a different temperature or a different grind size will completely change the flavor profile of a brew, even if the average extraction yield is the same. We can therefore expect that the choice of using agitation or not will have a dramatic impact on the flavor profile, even though it is a more indirect one.

Hence, *agitation is a process that increases extraction.* This is true because inside the slurry, agitation creates turbulent flows that can be characterized by a complex network of flow patterns at all scales. The Rayleigh-Taylor instability illustration, shown earlier, is one example of turbulent flow: if you were to zoom in on any part of the illustration, you would reveal new, similarly shaped complex patterns, much like fractals in mathematics. In practice, this means water flows efficiently between all coffee particles of any size where the coffee bed is agitated, ensuring that no single coffee particle is left underextracted. Further, turbulence drastically enhances how fast the chemicals move away from the particles and onto the slurry. Both effects make the extraction more aggressive and more even across the agitated part of the coffee bed.

The Melodrip breaks up a kettle stream, forcing it into the dripping regime and drastically reducing the amount of agitation that is imparted to the slurry. Photo by Dylan Siemens and Elika Liftee

In the context of manual pourover coffee, agitation will rarely reach the deepest levels of a coffee bed; only its upper layers will get agitated and extracted more evenly. Agitation does not impart only benefits, however. As you can imagine, lifting the coffee particles can also increase the migration of fines, potentially clogging the filter or reducing the clarity of a brew. In my experience, these effects happened only with the most extreme cases of agitation combined with channeling (e.g., by using a wide-mouth kettle) or in the presence of a large amount of fines. Therefore, maximizing agitation seems to produce desirable effects in scenarios where a grinder and coffee produce a relatively uniform particle size distribution, combined with thick filters that have a good load capacity. (See Chapter 5 for more details about load capacity.)

6.4 The Impact of the Water Stream on the Slurry

Studying how a stream of water impacts a "pool" of still water (in our case, the coffee slurry) is important to understanding how we can control agitation. A key point that determines how much turbulence is imparted to the slurry is the shape of the water stream—specifically, its irregularities—at the point where it hits the surface of the slurry. A smooth kettle stream will plunge inside the slurry almost undisturbed and create a strong water current that reaches deep inside the slurry, with little turbulence. This creates more currents of water that rise from the bottom to the surface around the central plunging point, lifting some coffee particles without necessarily separating them from each other. The momentum carried by the strong downward current might even dig a hollow in the coffee bed, which can later be problematic unless it is washed out with a swirl.

Detailed physical models of hollows caused by plunging water jets have been built by scientists to improve the safety and durability of hydroelectric power dams (e.g., Bollaert 2002; Bollaert and Schleiss 2003). Some of the conclusions from these studies are intuitive: thicker and faster streams at the impact point with the slurry will carry more momentum down to the bottom and will dig a larger hollow. However, these models contain another key insight that is not as immediately intuitive: a stream that entraps more air as it impacts the water surface will lose a lot more momentum before it reaches the bottom, thereby reducing the size of the hollow (e.g., see Duarte, Pinheiro, and Schleiss 2016). This happens because the air bubbles that are carried down are less dense than water, causing them to slow down because of buoyancy forces pushing them upward.

Friction between the downward-flowing water and the air bubbles will carry the bubbles downward, against buoyancy, for a little while; this efficiently slows the stream of water, making it more turbulent but also restricting its depth of reach into the slurry. As the stream of water reaches farther down, it slowly dissipates because of the friction forces—both with the air bubbles and the surrounding still water—until it becomes so weak that the buoyancy force acting on the air bubbles succeeds in pushing them back up. The depth at which air bubbles penetrate down into the slurry is therefore a useful visual indication of how far down we managed to create a significantly turbulent flow. In our previous discussion, we explained that this turbulent flow is what we call *agitation* in the context of coffee brewing; it is desirable because it can lift coffee particles in a chaotic way that is efficient at separating particles from each other and make extraction more even.

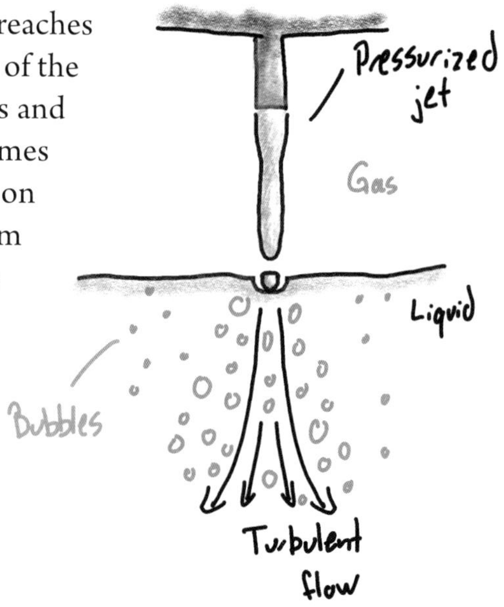

Schematic of a plunging jet reactor. A pressurized jet of fluid is released in a still pool to drag in surrounding gas and enhance chemical reactions between the fluid and gas.

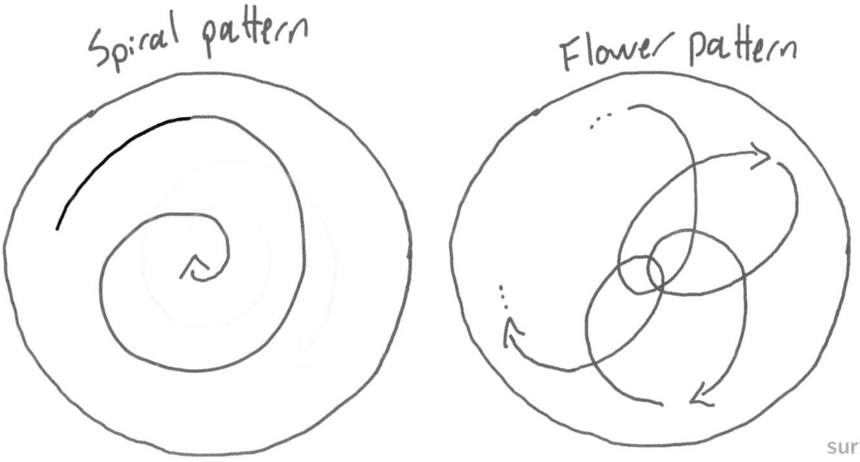

Example pouring patterns that can help achieve even surface coverage

Turbulent agitation can also dig a hollow in the coffee bed, but the hollow tends to be wider, and it is easier to fix with either a swirl or by distributing the stream of water around the surface of the slurry. Many different pour patterns have been adopted in the coffee community; I don't believe the exact pattern has an important impact on coffee preparation, as long as the water is distributed relatively evenly across the surface and the accelerations are not too sudden, to avoid breaking up the stream. I therefore recommend a pouring pattern that is as easy as possible to mindlessly repeat while evenly covering the surface. I personally like to use either a flower or a spiral pattern.

We have now set up a goal to achieve with our kettle stream: we want to entrap some air where the stream plunges into the slurry to maximize turbulence and agitate the coffee bed, and we want to control how deep this turbulence goes. Another unexpected field of science can help us with this: the field of plunging jet reactors. The main goal of these chemical reactors is to get a liquid to react chemically with a gas as fast as possible. This is hard to achieve because liquids and gases tend to have very different mass densities, and they therefore split into two layers that touch only along a small surface. A jet of fluid that plunges in the still liquid in a controlled way can carry some gas bubbles into the still pool, drastically increasing the surface of contact and therefore maximizing the rate of chemical reactions (e.g., Haraldsson 2000; McKeogh and Ervine 1981; Yin et al. 2018). Because air bubbles also make the stream of water turbulent, scientists researching this topic have been optimizing precisely the same parameter that we want to optimize for coffee brewing!

Qu et al. (2011) gathered experimental data and built a detailed physical model of plunging jet impacts to determine, among other things, how deep the air bubbles reach in a pool of water. One very interesting fact their study uncov-

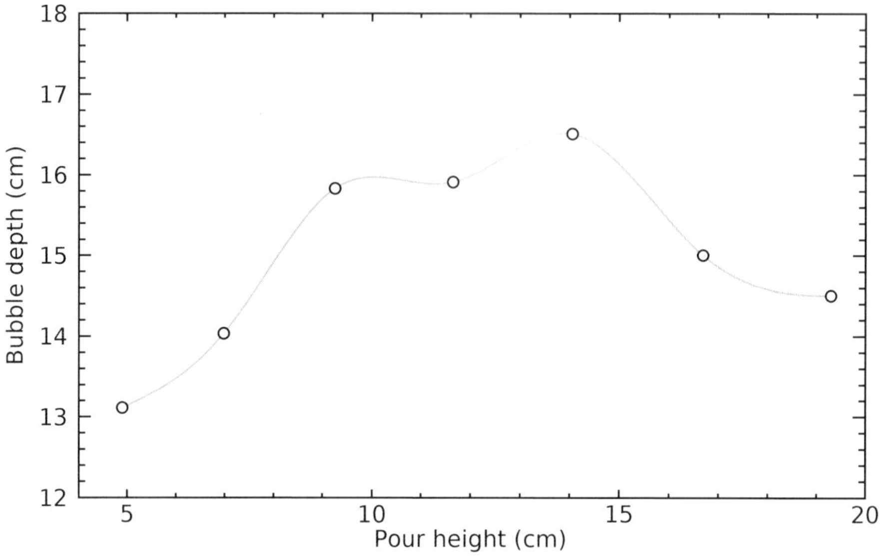

Depth of air bubbles imparted by a plunging jet on a still pool of water as a function of pour height, measured by Qu et al. (2011). A maximum depth happens when the breakup length of the jet is slightly below the surface of the pool; lower pours will cause less overall agitation, and higher pours will cause a shallower agitation depth.

ered is that the agitation does not keep getting deeper as we pour from higher up. Instead, for a given gooseneck and flow rate, *there is a specific height that maximizes the depth of agitation*, and pouring from any higher or lower will either decrease agitation or concentrate it closer to the surface of the slurry.

We can explain this with two distinct mechanisms at either end of the range in pour height. When we pour from a low height, the stream of water is very smooth when it hits the slurry, and almost no air gets entrapped. In this case, we create strong currents in the slurry, possibly digging a hollow in the coffee bed, without imparting a lot of turbulence. When we pour from too high, the stream of water will have had time to develop larger perturbations, and it will hit the surface in a more irregular form. When this happens, much more air will be entrapped when the water lands on the surface, because the rough shape of the stream will deform the surface a lot more, causing large pockets of air to be sent down into the slurry. This will drastically slow down the downward stream of water and will cause it to dissipate entirely in a very wide and shallow shape (McKeogh and Ervine 1981; Shonibare and Wardle 2015). In the most extreme cases, where the stream hits the slurry in the form of droplets, agitation will be significantly reduced (e.g., Guilizzoni, Santini, and Fest-Santini 2019).

In the context of coffee, the deepest agitation will occur when a kettle is placed just below the height where the stream would begin breaking up as it hits the surface of the slurry. It is easy to notice because this is when you begin to hear a splashing sound. To identify the pour height that maximizes agitation, it is therefore useful to first choose a pour rate that suits you, begin pouring from a low height, slowly raise the kettle until you reach the point where you begin hearing some slight splattering sounds, and then lower the kettle slightly. This is the height you will want to pour from to achieve a maximal agitation of the slurry.

We now understand how to maximize agitation with a given kettle and flow rate, but we have not yet investigated how different kettle properties affect our ability to impart agitation into the slurry. Earlier, we saw that kettles with smoother gooseneck tips always have a longer breakup, and that those with wider goosenecks will have their longest breakup at faster flow rates. We also saw that each kettle has a specific flow rate that maximizes its breakup length. In all these cases, we can simply adjust the height of the kettle such that the breakup point is just below the surface of the slurry. This means we will pour from the highest point if we choose the flow rate that maximizes our specific kettle's breakup length, and we will pour from lower down if we choose a flow rate that is either faster or slower. Using kettles with smoother goosenecks will generally allow us to pour from higher up and achieve more turbulent agitation of the coffee bed.

However, even if we always adjust our kettle height such that the breakup length is just below the surface of the slurry, these different scenarios will result in different amounts of agitation. When we pour from higher up, the water will

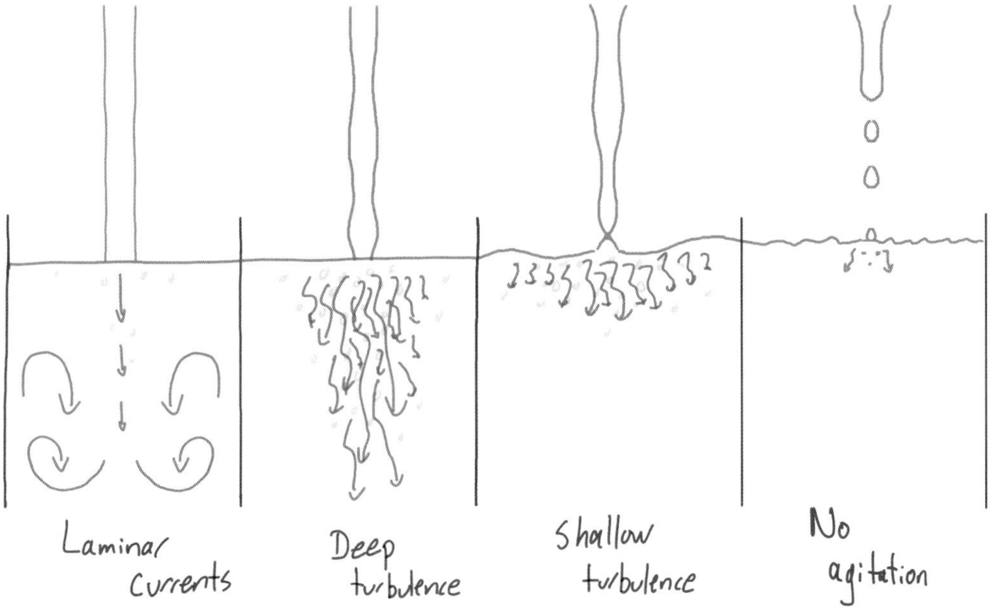

Types of agitation that result from different pour heights, with respect to the breakup length

have had more time to accelerate because of gravity, and it will impact the slurry at a faster velocity. This can be illustrated with the following equation:

$$v_{\text{land}} = \sqrt{v_{\text{kettle}} + 2gL}$$

where v_{land} is the velocity with which water lands on the slurry, v_{kettle} is the velocity as it exits the kettle, g is the gravitational acceleration on Earth (9.8 m/s^2), and L is the distance between the tip of the gooseneck and the surface of the slurry, which we call pour height for simplicity. For example, a pour height of 10 cm will contribute a considerable additional 1.4 meters per second to the water, compared with water poured from just above the slurry. Baylar and Emiroglu (2004) showed these effects with experimental data; the figure below displays their measured bubble depths for different roughness of streams and water landing velocities.

Therefore, if we compare two situations where the breakup point happens just below the surface of the slurry, the one with the higher kettle will cause more agitation. In other words, smoother kettle goosenecks generally allow a barista to cause more turbulence in the slurry, because their breakup point can be made to happen further down the water stream. At this point is important to understand a subtlety: I am not saying that pouring from higher up will always

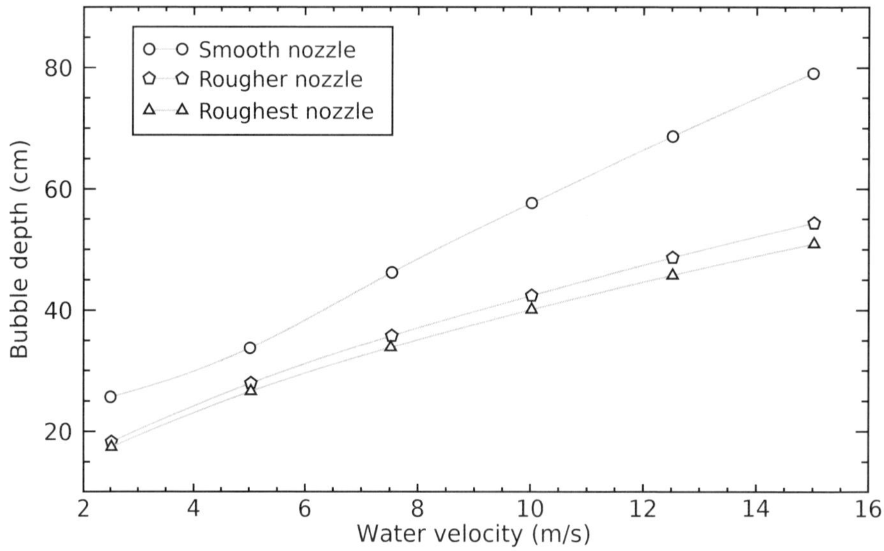

Experimental measurements from Baylar and Emiroglu (2004) showing that smoother water jet nozzles or faster pour velocities generate deeper agitation. All measurements were obtained with a fixed pour height. They used higher pressures to achieve faster water velocities at a fixed pour height, which we cannot do with typical pourover kettles.

maximize agitation! If you pour from so high that your kettle stream breaks up before it lands on the slurry, you will only be causing shallow agitation and therefore not affecting the coffee bed or extraction. You need a kettle stream that remains cohesive for longer in order to pour from a greater height *and* cause more agitation.

The Decent espresso machine pourover spray head

Kettles with narrower goosenecks will have their longest breakup point at a slower pour rate, and vice versa; selecting a kettle that suits your desired pour rate or using a flow restrictor[35] is therefore important. Selecting a thin gooseneck will allow you to extend your brew time and increase the amount of time during which the slurry is being actively agitated. Slower pour rates will also be easier to keep stable and facilitate replicating brews.[36] Conversely, a wider gooseneck will cause deeper agitation at its optimum pour rate and height.

A most extreme example of small gooseneck openings could be achieved by having more than one gooseneck on a single kettle or installing a spray head at the end of a gooseneck to produce several thin streams. While this could work in principle, we would quickly be faced with a fundamental limitation of conventional kettles: they all rely on the weight of the water to push itself out of the kettle. Dividing out this force that gravity affords us across several openings would cause a flow so weak that it would belong to the dripping regime, preventing any agitation—much like Melodrip brews. A solution to this problem could be to invent a dangerous kettle, where the water is shot out from a pressurized container. (What could go wrong?)

The Decent espresso machine ingeniously went around this problem by using a spray head adapter to its espresso portafilter, allowing it to basically become a pressurized kettle. In that scenario, the pressure drop can be much stronger than gravity, and dividing it among many streams becomes a possibility. This likely explains why Rao (2020, Feb. 4) observed higher extraction

[35] These are small devices that can be inserted into the entrance of the gooseneck inside the kettle. Flow restrictors installed on the outside tip of a gooseneck such as the Mandritech Hand Drip Water Valve Set can drastically reduce its breakup length and should thus generally be avoided.

[36] Some pourover kettles have a counterweight in their handles, making stable pours even easier.

yields when using the Decent spray head compared with manual pourovers; the pressurized streams allow us to impart a deep and more even agitation across the surface of a slurry. Following an empirical relation developed by Qu et al. (2011), the depth of agitation d_{ag} is proportional to the water velocity when it lands on the slurry v_{land} and on the hole size s_0,

$$d_{ag} \propto (v_{land})^{0.78} (s_0)^{0.67}$$

showing that the landing velocity has a slightly larger impact than the hole size. As a consequence of this relation, distributing the same discharge across more holes will cause agitation to reach slightly deeper with a larger number of holes N or a faster discharge Q:

$$d_{ag} \propto Q^{0.58} N^{0.23}$$

In other words, a 10-hole spray head would agitate the slurry 27% deeper than a 3-hole spray head, with the same discharge. However, another consideration limits the optimum number of water streams: heat loss. If the flow is divided across too many streams, each of them will be very thin, and they will lose heat faster to the surrounding air during their fall, because their surface per unit volume exposed to air will be larger. In the context of pressurized streams, it is also interesting to note that the height between the spray head and the slurry does not offer a benefit in terms of the landing velocity, because the gravitational acceleration along the fall will usually be negligible compared with the initial velocity of the stream. Hence, considerations of breakup length and heat loss become more important in this situation.

This situation of pressurized streams distributed evenly across the slurry

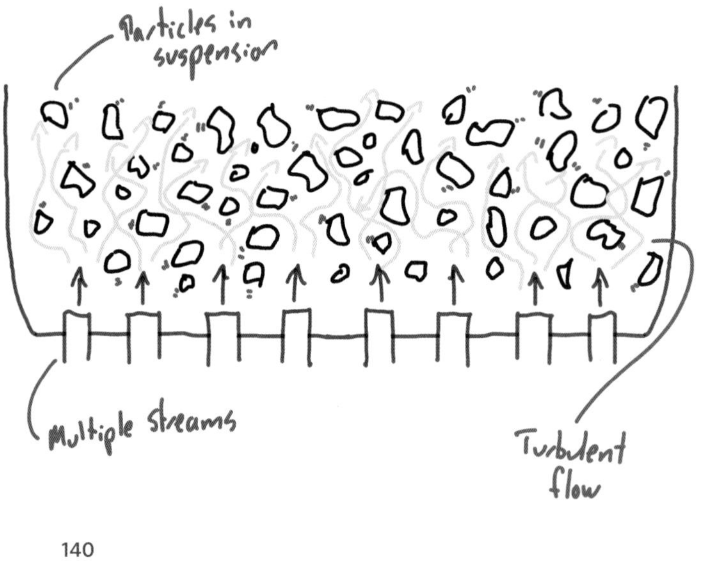

Illustration of a fluidized bed, where a large number of narrow, pressurized streams of gas or liquid agitate a porous medium to the point where it begins to behave like a fluid

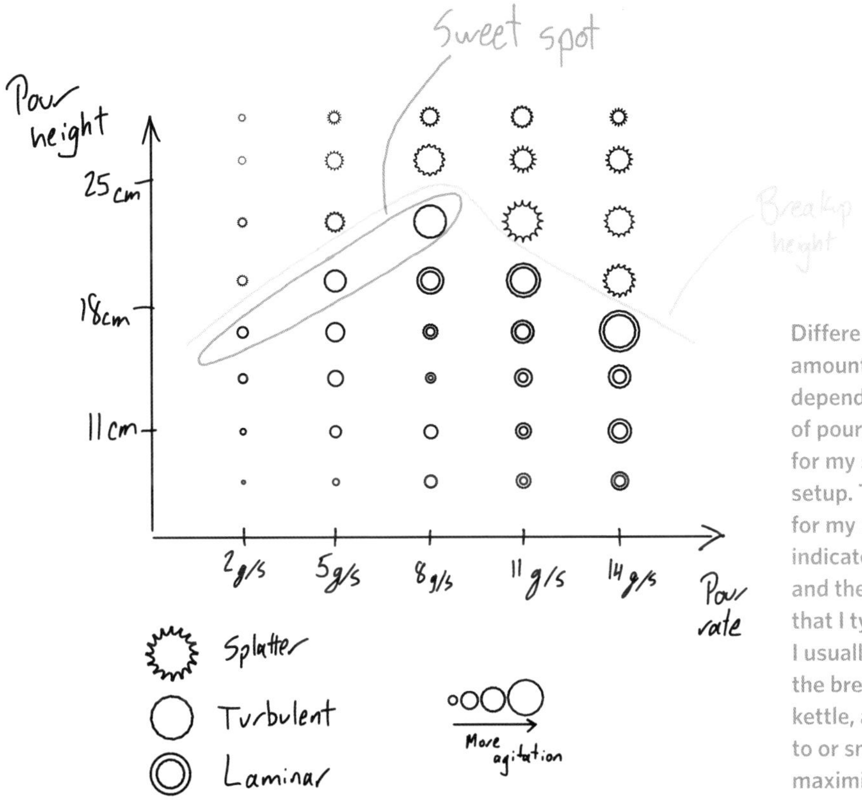

Different regimes and amounts of slurry agitation, depending on combinations of pour rate and pour height, for my specific pourover setup. The breakup length for my Stagg EKG kettle is indicated as a full gray line, and the "sweet spot" region that I typically use is circled. I usually pour slightly below the breakup length of my kettle, at a pour rate equal to or smaller than one that maximizes breakup length.

is similar to the concept of fluidized beds, widely used in chemical engineering (Crowe 2005). This phenomenon can even be applied rather spectacularly to transform a pool of sand into a fluid and even allow one to swim in it as if it were water (Rober 2017). This is potentially a great technology for coffee brewing because it would make extraction more even. Fluidizing the entirety of a coffee bed could, however, make it a less efficient medium of filtration, depending on how fast it settles down when the pressurized streams are shut off. Regardless, this could also worsen fines migration and lead to filter clogging with all but the most uniform particle size distributions. All of these downsides could be mitigated either by using a thick and efficient filter or by adding a nonextracting porous medium, such as washed sand, below the bed of coffee.

To summarize what we have learned in this section, I built a map of example pour rates and pour heights for my particular setup (i.e., a 22-gram dose in a V60 with a Stagg EKG kettle), shown above. The "sweet spot" corresponds to the range of pour heights and rates that I use, depending on how prone to clogging a coffee is. A simplified version of the map is shown on the following page, where we focus on the "sweet spot".

The last map, also on the following page, shows specific problems that may arise from using different combinations of pour rates or pour heights.

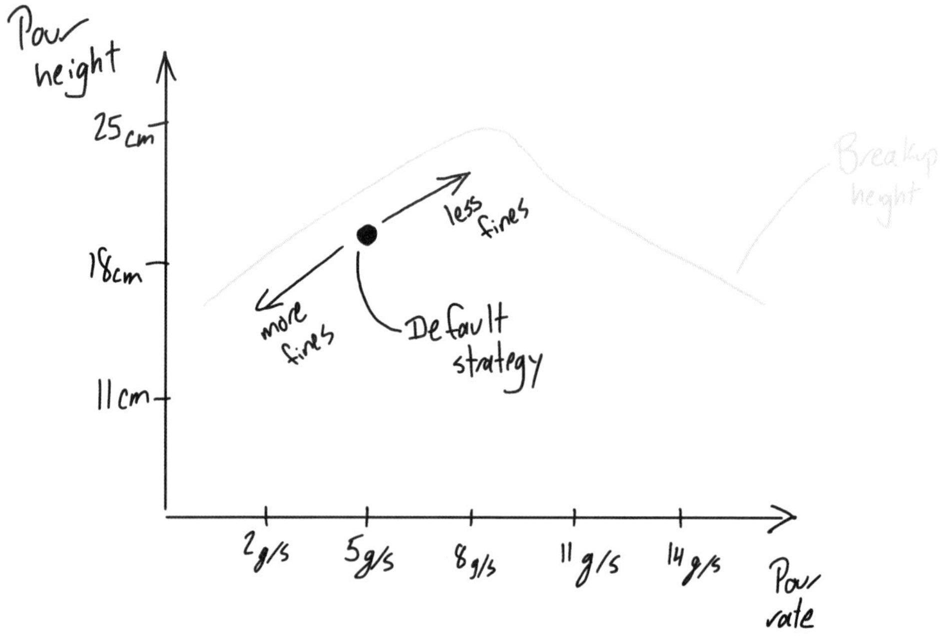

This map illustrates how I adapt my pour strategy within the "sweet spot," depending on how much fines a coffee generates. Coffees with more fines are more prone to filter clogging, and I therefore use slightly less agitation when brewing them.

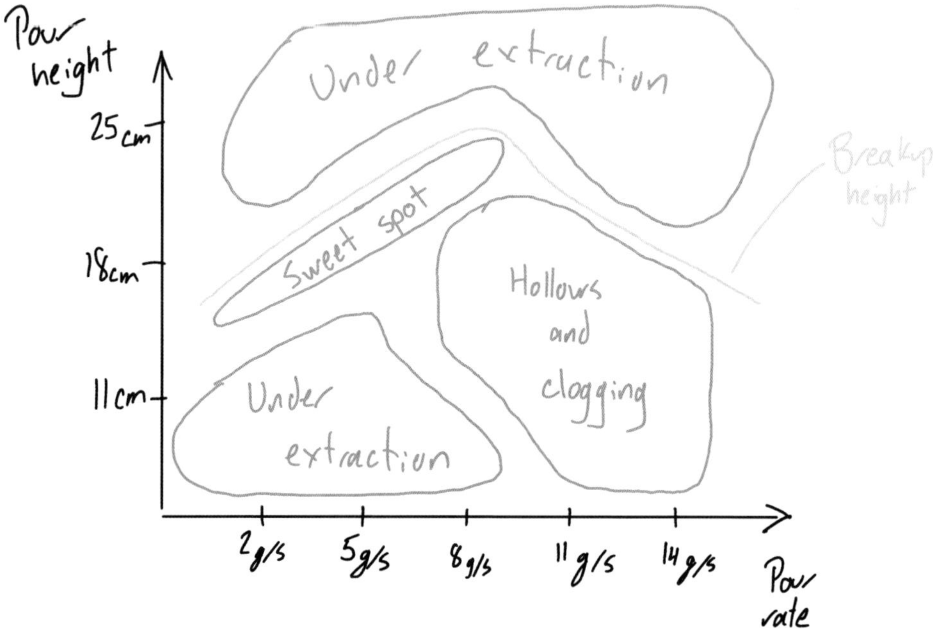

An illustration of the different problems that may arise when choosing pour rates or pour heights very different from the "sweet spot" region

6.5 Insulation

One further important point about kettles is their ability to maintain a stable temperature during coffee preparation. While most kettles offer an option to maintain a desired custom temperature, some require pressing a button every time the kettle is placed back on its base, without which the kettle will let the water cool down freely. I recommend using a meat thermometer in the kettle, at least once, to assess how reliable its temperature indicator is and how stable the temperature remains after removal of the kettle from its base. The first few degrees around 100°C (212°F) will always be lost more quickly, because heat transfer is faster when the temperature difference between water and air is larger; this means that good insulation of the kettle is important when we are brewing at temperatures near the boiling point.

Generally, kettles with a larger water capacity retain heat better, because more heat needs to be transferred away from the larger mass of water to reduce its temperature by the same amount. This comes at the cost of pour stability and repeatability of pour rates, however, as heavier kettles are harder to control. Some manufacturers (e.g., Brewcoat) offer heat-insulating coatings that can fit a wide variety of kettles and improve their temperature stability, which is especially useful for low-capacity precision kettles. I found that using a thin layer of

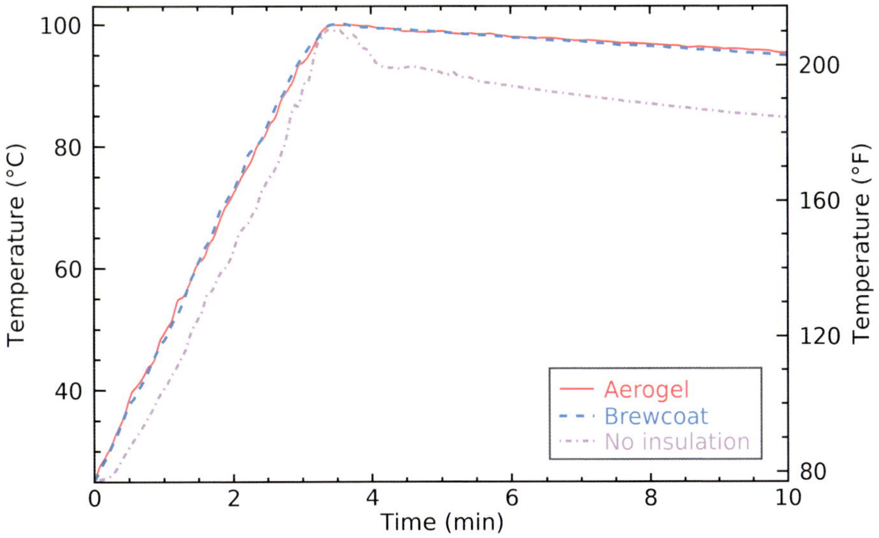

Efficiency of water heating and insulation for a 0.6-L Brewista Artisan kettle with different coatings. Adding any insulation reduces the energy needed to reach the desired temperature and makes it more stable, especially at temperatures above 95°C (203°F).

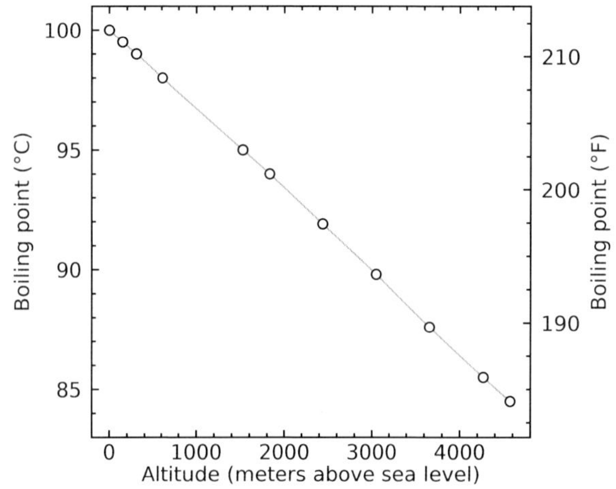

Temperature at which water boils, for different altitudes. Data from the National Oceanic and Atmospheric Administration

Altitude		Boiling point	
(m)	(feet)	(°C)	(°F)
0	0	100.0	212.0
150	500	99.5	211.1
305	1,000	99.0	210.2
610	2,000	98.0	208.4
1,524	5,000	95.0	203.0
1,829	6,000	94.0	201.1
2,438	8,000	91.9	97.4
3,048	10,000	89.8	193.6
3,658	12,000	87.6	189.8
4,267	14,000	85.5	185.9
4,572	15,000	84.5	184.1

aerogel blankets[37] can achieve the same results, but they tend to leave a white powder on the fingers when they are manipulated.

One additional consideration to keep in mind when selecting a brew temperature is the local altitude at which the coffee is being prepared. Because water boils at a lower temperature at higher altitudes (due to the lower atmospheric pressure), kettles should not be set to a temperature higher than the local boiling point. The figure and table above provide boiling point values from the National Oceanic and Atmospheric Administration.

We will discuss the considerations of selecting a kettle temperature in chapters 7 and 10, in the context of dripper properties and brew techniques. We will see that *the slurry temperature is generally much lower than the kettle temperature,* depending on brewers. This means that it is important to consider heat loss caused by the brewer, not only by the kettle, when selecting a brew temperature.

[37] Aerogel blankets are flexible blankets injected with fragments of aerogel, an ultralight and porous material that has an extremely low thermal conductivity and other unusual properties.

CHAPTER 7

Drippers

Over the years, several manual filter coffee drippers with a wide variety of shapes, materials, and features have been developed. Some, such as the V60, AeroPress, and Kalita Wave, have become wildly popular, but the reasons behind their wide adoption and how they affect coffee brewing can seem quite obscure. In this chapter, we investigate how the most important properties of coffee drippers affect brewing, and we apply this understanding to assess the properties of currently available drippers.

7.1 Effective Filter Surface

Paper filters are widely used across manual coffee drippers and, as we explored in Chapter 5, they play an important role in keeping the coffee bed together while allowing a stable and uniform flow of water. The design of a dripper can interact significantly with the filter performance, by changing the effective area through which water can flow. For example, a filter sticking to a perforated smooth bottom will allow water to flow through only where the dripper holes are located. This can dramatically reduce the effective filter surface, slow down the flow of water, and make the filter much more prone to clogging because it reduces its load capacity.[38] One solution to increase the effective filter surface is to add ridges at the bottom of a dripper, or even a coarse tea strainer mesh between the filter and the surface; this will lift up the filter and allow water to pass through its complete surface before making its way to the exit holes of the dripper.

[38] Remember from Chapter 5 that the load capacity corresponds to the quantity of fine coffee particles a filter can absorb before it clogs.

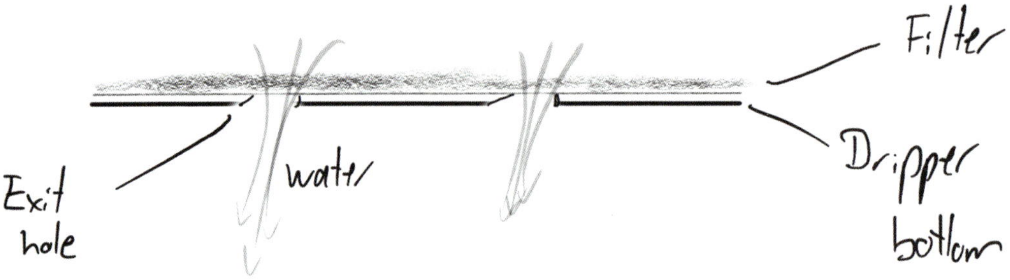

Schematic of a coffee filter that adheres to the bottom of a dripper except for the dripper holes, greatly reducing its effective surface area where water can flow through the filter

7.2 Geometry

One of the most important and immediately apparent features of coffee drippers is their geometry. The geometries of the most widely used drippers fall into either of two categories: flat-bottom or conical. Conical drippers, used in combination with cone-shaped filters, make small coffee doses deeper, compared with cylindrical drippers. In addition to this, the change in coffee bed depth, past a coffee dose of about 10 grams, is less pronounced as a barista changes the coffee dose slightly; this makes the dripper more versatile.

As we discussed in Chapter 4, in an idealized scenario where no agitation occurs and no water bypasses around the coffee bed, Moroney et al. (2019) determined that the spatial uniformity of extraction is different in conical ver-

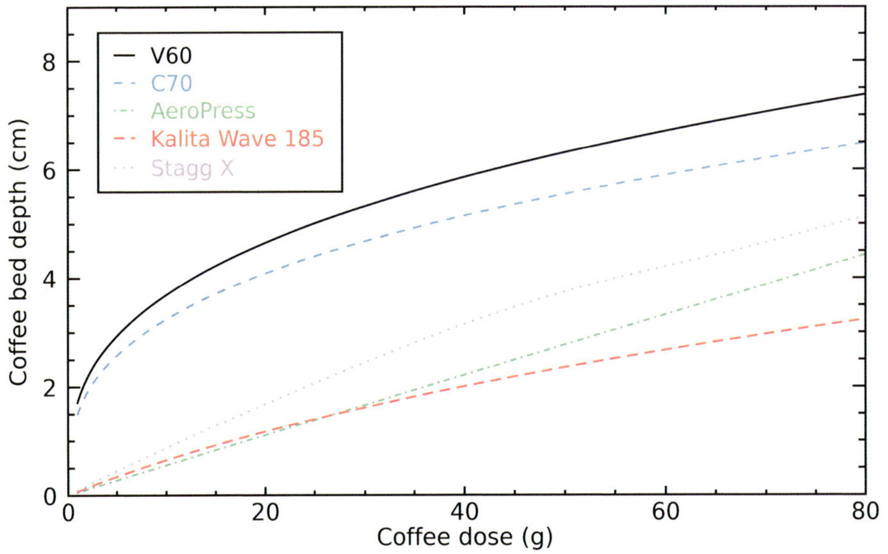

How bed depth varies with coffee dose across different dripper types. For larger doses, drippers with flat bottoms perform better than conical drippers.

sus flat-bottom drippers. Flat-bottom drippers tend to have extractions that are more uniform laterally but much less uniform across the depth direction because the bottom of the coffee bed always encounters more concentrated water. This latter effect is partly compensated for in a conical dripper because more water will pass by each coffee particle, near the bottom.

However, any conical dripper must make a compromise; if its walls are smooth and allow the filter to stick to them completely, the effective filter surface that is aligned with the dripping hole is very small compared with the total volume of coffee, making it prone to clogging. One way to mitigate this to add ridges along the filter walls to drastically improve filter surface, but that will cause some water to flow through the filter before it passed through the entirety of the coffee bed, an effect that is often called *bypass* in the coffee community.

7.3 Flow and Bypass

The flow through a dripper is usually limited by the number of holes at its bottom and whether a paper filter directly adheres to its surface. As we previously discussed, the effective filter area is drastically reduced when the filter adheres to the dripper, reducing flow and potentially causing the filter to clog. Adding ridges at the bottom of a dripper is therefore important to lift the filter away from the holes and ensure a maximum effective filter area, in which case the flow through the dripper is usually most constrained by either the coffee grind size, or the bottom holes of the dripper if they are very small.

Conical drippers that use ridges to lift the filter along the walls can also make the flow faster by allowing a fraction of the water to bypass completely or in part around the coffee bed. This can also happen to a lesser extent in flat-bottom drippers with prepleated filters if the column of water is tall above the coffee bed and the filter walls do not adhere to the dripper walls. When water completely bypasses the coffee bed, the effect on the resulting cup of coffee will be to reduce its concentration. This bypass effect will potentially also reduce the average extraction yield because less water was used as an actual solvent to extract the coffee solubles during the percolation phase. In my opinion, such water bypassing completely around the coffee bed is not necessarily a problem if you can still obtain your desired combination of average extraction yield and coffee concentration. Such "complete" bypass will tend to happen more commonly in a conical dripper when the column of water above the coffee bed is tall.

One more important issue arises when water partially bypasses the coffee bed. This can happen in conical-shaped drippers that have ridges along their walls, such as the V60, especially if the coffee bed has a high hydraulic resistance. In that scenario, some water passes through the upper layers of the coffee bed, exits halfway through the filter along the dripper wall, and makes its way to the dripper hole by flowing between the ridges. This causes the bottom layers of the coffee bed to be underextracted compared with upper layers and can give rise to a different taste profile as well as a lower average extraction yield. I suspect this is the major reason coffee prepared in flat-bottom drippers seems to result in slightly higher average extraction yields and a higher perceived sweetness.

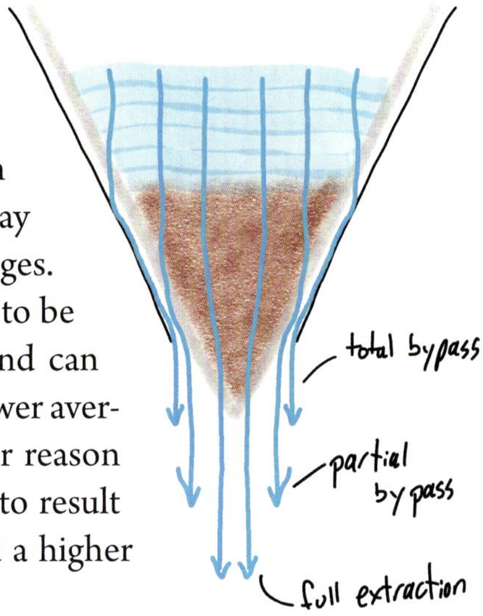

Bypass can be observed with a transparent plastic V60, especially when the dripper is filled with water, but it is very hard to tell by eye how much water is bypassing or where, exactly, it crosses the filter. The data that finally convinced me that ridged conical drippers such as the V60 can have a

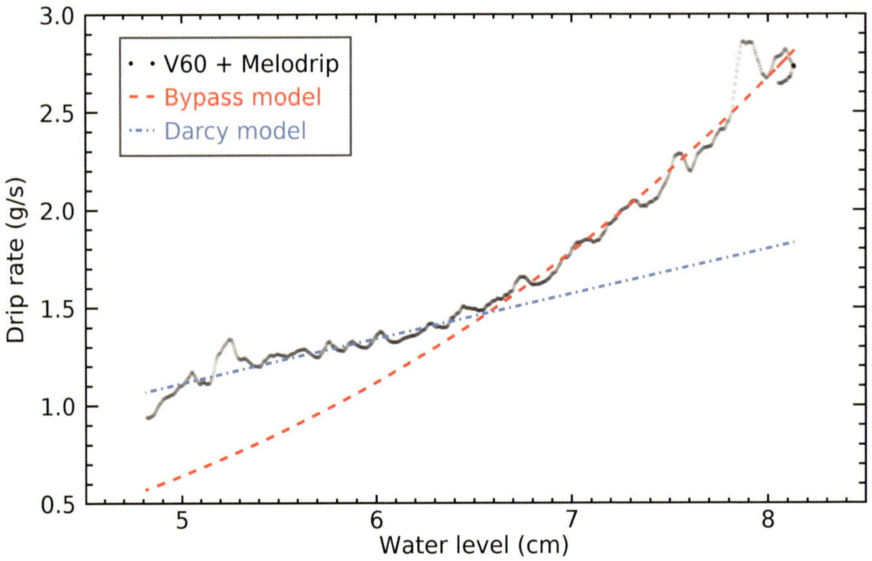

Drip rate as a function of water height immediately following the last pour in an agitation-free V60 brew. The flow of water became much faster than the predictions from Darcy's law when the water level rose above 6 cm, in a way that matches a model of a conical filter flowing freely without dripper walls. The drip rate followed Darcy's law when the water level was low, indicating that a negligible amount of bypass was taking place in this situation. I did not observe this effect consistently in all my V60 brews; bypass seems to get triggered at random, probably because it depends on the weight of water, the hydraulic resistance of the coffee bed, and how the filter was positioned in the dripper and prerinsed.

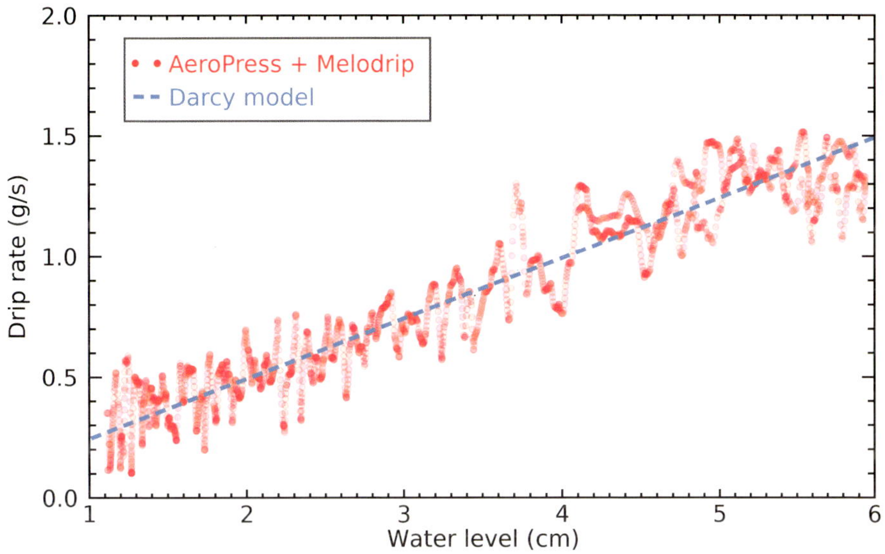

Drip rate as a function of water height in the bottom part of an AeroPress used as a gravity-driven dripper. I used a Melodrip to remove as much agitation of the coffee bed as possible. The flow fluctuates but globally follows Darcy's law, which predicts a linear relation between drip rate and water height.

significant amount of bypass came from logging my beverage weight and total weight during several pourover brews with the setup I will describe in Chapter 11. I observed that Darcy's law, adapted for conical drippers, sometimes did not fit my data at all, even in the complete absence of agitation. Darcy's law[39] predicts that the drip rate should depend linearly on the height of water in the V60. In reality, the drip rate becomes much faster than this prediction as the water column grows taller.

This bypass of water around the coffee does not happen in a predictable way, and the flow of bypass water can change significantly during a brew or between brews as channels of bypass water connect and disconnect outside the filter and between the ridges of the dripper wall. It tends to be less important if the filter adheres more to the dripper walls—for example, if the filter was rinsed thoroughly with pressurized tap water and not subsequently displaced. Note that pouring water at the center of the slurry only does not reduce bypass by any means.

I verified that this effect does not occur when using the bottom part of an AeroPress as a gravity-driven dripper. The figure above shows that the drip rate followed predictions of Darcy's law based on the height of the water column, indicating that no water was bypassing around the coffee bed.

---

[39] adapted to a conical dripper

7.4 Thermal Insulation

One of the biggest impacts of a dripper is its capacity to thermally insulate the slurry during a brew. Two material properties are relevant to this discussion: heat capacity[40] and thermal conductivity. Before we explain them, it might be useful to get rid of some confused intuitions we may hold from daily experience with respect to the physics of heat transfer.

When we touch an object and feel that it is cool or warm, we are not actually measuring its temperature. Instead, we are measuring how fast it is transferring heat to or away from our skin. Hot objects will tend to transfer heat our way much faster, but this also depends on how thermally conductive the material is. More conductive objects will transfer heat faster, maximizing our perception of the temperature difference between our skin and the object.

One of the fundamental principles of physics is that two objects in contact with each other will eventually come to have the same temperature. This happens faster when the objects are more thermally conductive, because they can exchange heat faster. In slowly moving toward this equilibrium, heat always flows, on average, from the warmer to the cooler object. Therefore, if we leave a piece of heat-resistant rubber and a piece of metal in the oven for hours, we know they will come to the same temperature. But you can easily imagine which one would feel much hotter to the touch! Metal would burn you much more badly because it is highly conductive, even if it has the same temperature as the rubber. The same holds true with a freezer; you have probably noticed that metallic objects in the freezer feel much colder than objects made of plastic.

Now that we better understand thermal conductivity, let's talk about heat capacity. Another aspect of heat transfer may seem surprising: not all materials need to receive the same amount of heat for their temperature to rise by the same amount. Materials made of larger molecules typically must store more heat than those made of smaller molecules for the same temperature rise, and metals tend to require a very small amount of heat.[41] Heavier objects also generally have a larger heat capacity, simply because there are more molecules to be heated.

These clarifications can help us understand how the material from which a

---

[40] The heat capacity of a material is the amount of heat that is required to increase its temperature by 1 degree.

[41] The reason for this is related to electrons behaving in a special way in metals; they move around much more easily than in other materials, and this causes the electrons to be very efficient at transporting vibrations, motion, and electricity far away, making the materials both highly thermally and electrically conductive.

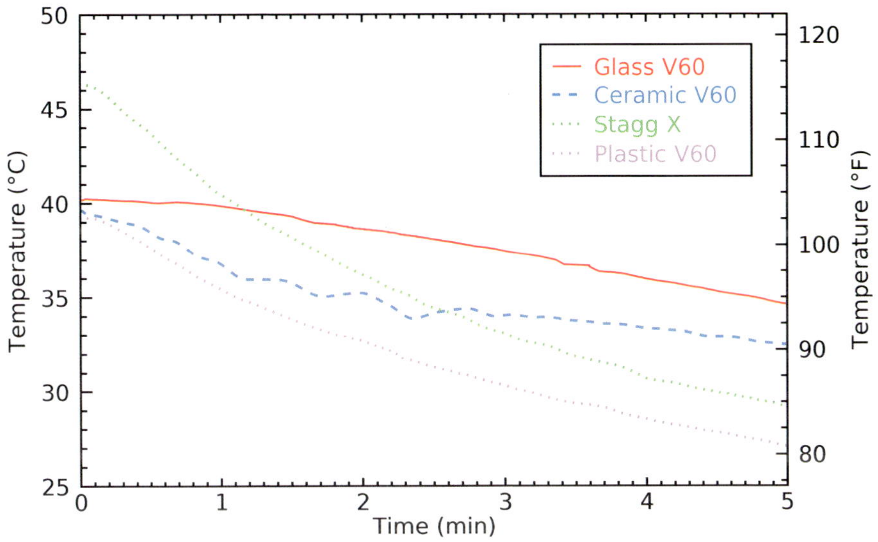

Temperature of empty drippers after preheating them with 99°C (210°F) water at an 8 g/s pour rate for 15 seconds. Drippers with less heat capacity will cool down faster after preheating.

dripper is constructed affects our slurry temperature. When we pour hot water into the dripper, the dripper will initially have a temperature much lower than the water, even if it was preheated. The figure above shows how short-lived the effect of preheating a dripper can be, depending on its heat capacity.

When hotter water is poured into the cooler dripper, heat will immediately begin transferring from the water into the dripper walls until the dripper reaches the same temperature as the water. If the dripper is heavy and has a large heat capacity, it will need to drain a lot more heat from the water before it reaches this equilibrium, and both the water and the dripper will stabilize at lower temperatures. This stabilization can happen almost immediately for very conductive materials, such as metal.

Once the temperature balance has been reached, the more-conductive drippers will continue transferring heat away to the surrounding air by conduction. This transfer, while generally slow (because air is not very conductive), is more important for highly conductive drippers. This transfer will again steal away more heat from the slurry. Some heat will also be lost as water evaporates at the surface of the slurry—but I found that using an insulating lid barely makes a difference because it needs to be taken off every time water is poured in.

There are two generally good ways to get a sense of how much heat a dripper will remove from your slurry. The heavier drippers, particularly those that become very hot to the touch during a brew, are indeed stealing and redirecting a lot of heat away from the slurry. The worst ones in that regard are ceramic and glass, as they are both heavy and conductive. Metal drippers are conductive but very light, so they don't steal as much heat, especially at the start of a brew. Plastic drippers are generally light *and* thermally insulating, and they will usually result in the highest slurry temperatures. In rare cases, vacuum-insulated dripper walls provide an even better heat insulation. This could seem counterintuitive, as we saw that the plastic V60 and vacuum-walled Stagg drippers lose their preheating benefits faster, but even at room temperature they end up stealing less heat from the water, compared with preheated ceramic or glass drippers.

It may sound like I'm saying glass or ceramic drippers are always bad, but this is not necessarily the case. Just keep in mind that (1) the kettle temperature does *not* reflect the temperature at which you are actually brewing, and (2) the least-insulating materials will cause much wider swings in temperature between pours. I personally prefer the taste of brews made with stable and hot temperature profiles, for all but the darker roasts.

I often see coffee professionals recommend the same brew temperatures for methods such as AeroPress and V60. This sounds weird to me because AeroPress is very insulating and will thus brew almost exactly at your kettle temperature, especially with the plunger on, whereas V60 brews will immediately lose about 5°C (9°F) just as your stream of water falls onto the slurry. Using addi-

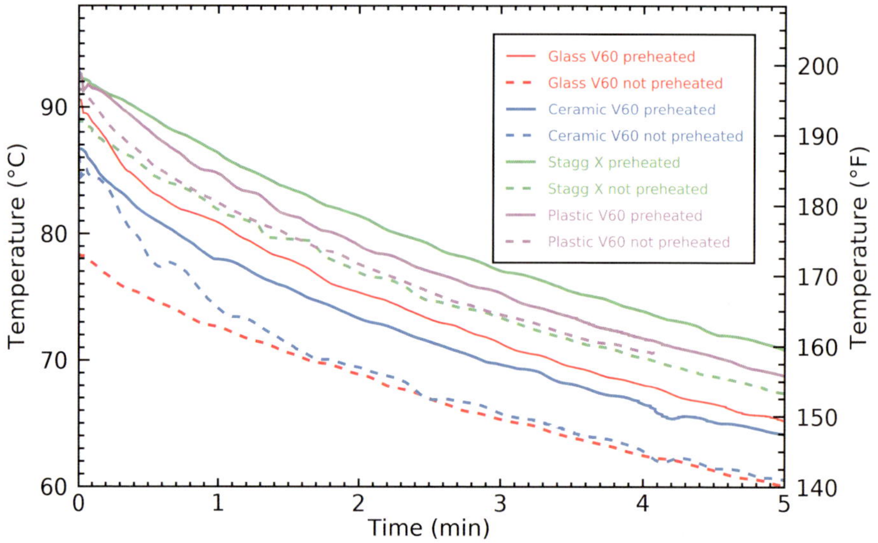

Temperature of 100 grams of water from a 99°C (210°F) kettle placed in drippers made of different materials. Solid lines indicate drippers that were preheated as described before, whereas dashed lines indicate the drippers were used at room temperature (22.5°C; 72.5°F). Drippers with more insulation and less heat capacity make the slurry temperature higher and significantly more stable over time.

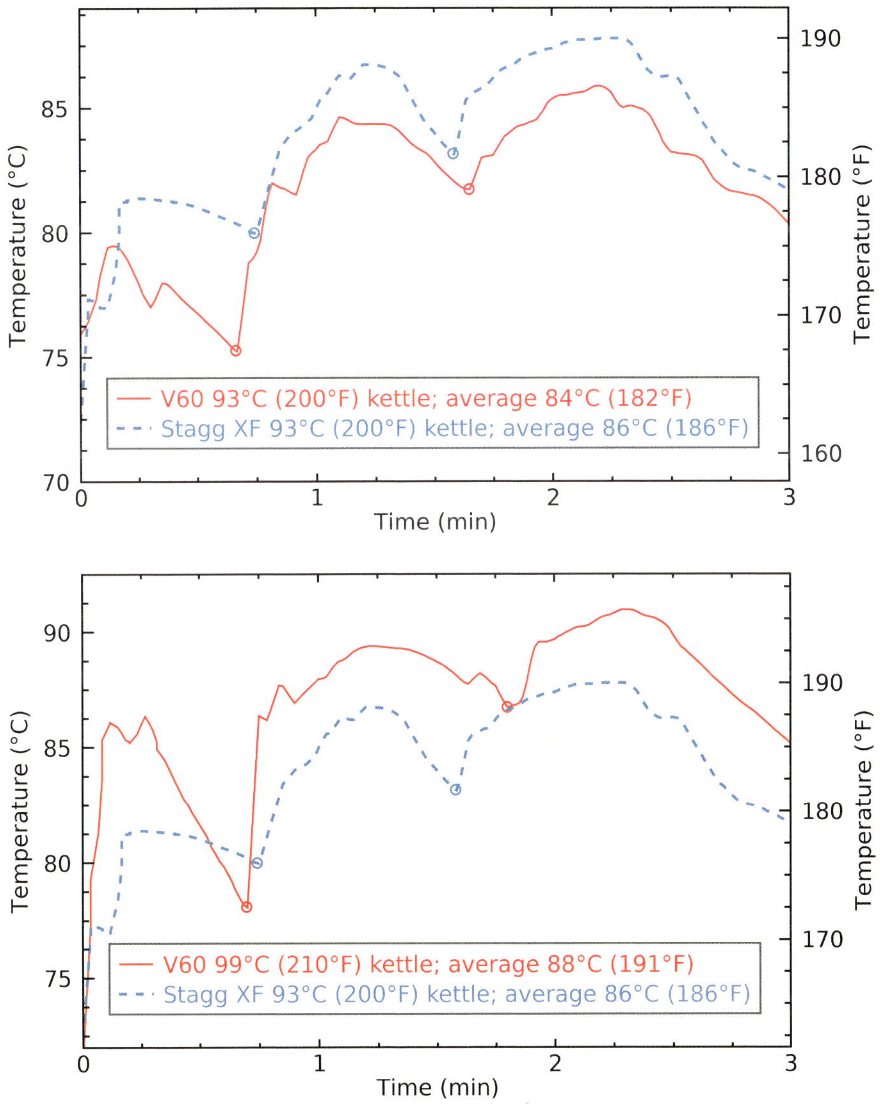

Slurry temperatures during pourovers made with a 45-second bloom and two subsequent kettle pours. Empty circles indicate the beginning of pours. Top panel: the plastic V60 (red line) has a lower slurry temperature than the vacuum-insulated Stagg XF (dashed blue line) for the same kettle temperature. Bottom panel: overcompensating for this slurry temperature difference by using a lower kettle temperature with the Stagg XF results in a more stable temperature profile, despite the slightly cooler average slurry temperature.

tional tools that come into contact with your water, such as the Melodrip, will also lower your slurry temperature unless you carefully preheat them, placing them in a glass of hot water between pours.

In my experience, only the most insulating drippers, such as AeroPress, can be used to brew coffee at a slurry temperature so high that it affects the taste negatively and results in an acrid or bitter brew (above 96°C or 205°F). This is also true of the siphon because the constant source of heat can compensate for heat loss. I usually brew with water close to boiling (99°C or 210°F) when I use the V60 or any other manual pourover dripper. I avoid going to 100°C (212°F) because it can cause kettles with narrow goosenecks to have an unstable stream. In Chapter 10, I will discuss an exception to this general rule: when I brew coffee that is roasted darker.

7.5 Dripper-Server Seal

In unusual cases, a dripper can form a perfect seal with the coffee server on which it is placed. When this happens, the air trapped in the server has nowhere to escape, and the addition of liquid coffee dripping from the slurry will compress it and increase the air pressure in the server. This can rapidly slow down a brew by decreasing or even removing the atmospheric pressure drop that drives the flow of a percolation brew. It is useful to always be mindful of that possibility. The Chemex is particularly infamous for this effect because the dripper and server are only one piece. The only place where air can escape is therefore through the spout, which means that the flow can rapidly stop if the filter collapses in the spout. This effect compounds with the smaller exposed surface area of the Chemex filter.

7.6 Example Drippers

I list here some popular drippers along with their specific features and some of their particularities—which could be perceived as advantages or inconveniences, depending on the goals of a barista.

Hario V60

This small, cone-shaped dripper comes in a variety of different materials (plastic, glass, or ceramic) and has ridges along its 60°-angle walls.

Water can bypass significantly around the coffee bed, depending on the brew method, grind size, filter, and the exact thickness of the ridges. This bypass will potentially reduce average extraction yield and evenness of extraction.

- Ceramic and glass drippers have lower and less-stable slurry temperatures compared with the better-insulating plastic version.
- The cone shape makes the V60 more versatile in terms of optimal coffee doses.

- A wide range of paper filters are available, making this dripper yet more versatile.
- This dripper probably has the largest effective filter surface relative to the coffee dose, making it particularly good for trapping coffee fines. This can help in obtaining a cleaner beverage and avoiding filter clogging.
- The circular base of the V60 makes it easy to swirl the slurry.

Fellow Stagg X and XF

These drippers come in two sizes, the smaller Stagg X and the taller Stagg XF. They are made of vacuum-insulated double walls and have angled walls that are curved outward and a narrow, flat bottom. They are intended for use with prepleated flat-bottom paper filters, which are held up by shallow ridges at the bottom. The flow of water is constrained by ten small holes at the bottom of the dripper.

- These are the most thermally stable pourover drippers I have ever encountered. They allow brewing at higher slurry temperatures to obtain slightly higher average extraction yields, assuming the coffee was roasted light enough to avoid bitterness. Even for darker roasts, they offer a more stable slurry temperature.
- The flow of water is relatively fast, provided that the filters do not collapse on the dripper holes. Unfortunately, this can happen relatively easily if the barista is not careful because the ridges at the bottom are quite shallow. I recommend lifting the filter walls very slightly when prerinsing the filters to ensure a good flow in order to avoid this problem. Alternatively, placing a small, food-grade silicon ring or a cut-out tea strainer mesh at the bottom of the dripper can eliminate this problem.
- Prepleated filters come with their own challenges: they can trap some coffee particles within the folds, and it is prudent to avoid pouring water outside the filter.
- When water is poured inside the filter, bypass will be a relatively smaller issue because the dripper walls are steep. Partial bypass through the coffee bed probably does not happen much, for the same reason. I believe this is part of the reason why I obtain slightly higher average extraction yields and noticeably sweeter brews with this dripper.
- A prewetted V60 paper filter can be forced to fit inside the dripper, allowing for more space above the coffee bed and removing the problems associated with prepleated filters. However, the exact way in which the filter is placed will

generally affect the flow of water, depending on how it clogs the holes. In my experience, this leads to brews that are better on average but less repeatable. I have obtained very good results by combining this trick with use of a bottom, circular tea strainer mesh.

- The base of this dripper comes with a small, removable rubber pad. Leaving the rubber pad on makes it almost impossible to swirl the dripper. Removing it helps, but the dripper can remain awkward to swirl, depending on what it stands on. Placing the dripper with its rubber pad on a removable V60 base—for example, the Hario Olivewood base—completely remedies this problem and makes the dripper as easy to swirl as a V60.
- This dripper is slightly less versatile than conical drippers in terms of optimal dose sizes, because the bed depth will vary more when the dose is changed.

Chemex

This one-piece glass dripper is made of a conical dripper section fused with a glass server. The dripper has no ridges on its walls, with the result that the filter adheres to the glass and the effective filter area is very small.

- Water bypass is less of an issue because the filter sticks to the dripper walls, except where the spout is.
- Its very thick paper filters offer a better filtration of coffee oils and fines but a slower flow.
- It is easier to clog a Chemex because of the small filter area, despite the filter's thickness, especially if the filter collapses in the spout. If this happens, the seal between the dripper and the server can completely stop the flow of water, even in the absence of coffee fines.
- Glass is not a good thermal insulator, and therefore the slurry temperature will be lower and less stable.
- The cone shape makes this dripper more versatile in terms of optimal coffee doses.
- Swirling the Chemex is a bit awkward because it is fused with the server.
- The slower overall flow may require grinding coarser, compared with a V60.

Origami

This conical dripper is made of a thin ceramic shaped like a prepleated filter, with a bottom hole larger than the V60. It is in several ways similar to a ceramic V60, but it has slight differences.

- Water bypass is like that of a V60, except that prepleated filters can be made to adhere to the dripper walls a bit more, which probably reduces bypass.

- The hole at the bottom of the dripper is larger than that of a V60, which will make the flow generally faster.
- The base of the dripper is awkward to use and the dripper can slip off from the base easily, especially if you hope to swirl the slurry. This has led home baristas to design and 3D print custom dripper bases.
- Its light weight means the slurry temperature won't be as low as that of a ceramic V60. Ceramic is conductive, however, so the slurry temperature decreases faster than it would in a plastic dripper. If combined with a conical filter, the limited contact between the filter and wall will mitigate the heat loss but increase bypass.
- This dripper is versatile in terms of dose size, much like the V60.

Kalita Wave 155

This flat-bottom dripper is made of stainless-steel metal with three ridges and three holes at its bottom. Its walls are slightly slanted, and the dripper is designed for use with prepleated Kalita filters.

- It is less susceptible to water bypass because of its flat bottom and gentle wall slope.
- The three small bottom holes reduce the flow of water, compared with most other drippers.
- Stainless steel has a small heat capacity, which means it will initially reduce the slurry temperature only slightly, but it is also a great heat conductor, causing the slurry to lose heat faster. The prepleated filters mitigate this issue because they trap layers of less-conductive air between the folds.
- The Kalita paper filters are prepleated and come with their own issues, as discussed in the Stagg description above.
- This dripper is less versatile than the V60 in terms of optimal dose size because the bed depth will vary more when the dose is changed. This is mitigated, in part, by the angled walls.

C70

This dripper is similar to the V60, except for its steeper wall angle, at 70°. This means the optimal dose will be slightly larger and the range of acceptable doses will be a bit narrower.

Lilydrip Devices

These small, conical ceramic devices can be added to the bottom of a V60 dripper to change the shape of the coffee bed. In my experience, they do not affect the average extraction yield or the taste much at all, but they may render the bloom slightly more efficient, eliminating the need to place the dry coffee dose in a nest shape. Their installation in the dripper requires flipping the paper filter, which is cumbersome, and they must be preheated carefully, without which they will initially lower the slurry temperature because of their high heat capacity. They affect flow and bypass in different ways, depending on their angles and ridges, and they make pouring uniformly across the slurry surface a bit more straightforward because pouring in circles will do the trick.

CHAPTER 8

Freshness

Roasted coffee is an inherently unstable product, and this represents one of the significant hurdles in preparing good coffee repeatably. As you have probably experienced, the flavor of coffee changes as it ages, often in a not-too-pleasant way. In this chapter, we review some of the causes of these changes, the factors that most affect how fast coffee ages, and the practices we can adopt to better preserve coffee at its best taste. For more information, I recommend following the research of the Coffee Excellence Center of the Zurich University of Applied Sciences, which studies matters of coffee freshness.

8.1 Carbon Dioxide

When coffee is roasted, chemical reactions in the coffee beans produce a large amount of carbon dioxide (CO_2), most of which remains trapped inside the beans under pressure. The exact amount of CO_2 that remains trapped, and how pressurized it is, depend on the roast profile. Faster and darker roasts typically generate more CO_2, and the beans degas faster (Smrke et al. 2018). While CO_2 does not bear a direct impact on the taste of filter coffee, it can interact negatively with the process of extraction. It is also responsible for the generation of crema in espresso, which looks beautiful but does not necessarily taste good (Hoffmann 2009). As coffee ages, it slowly loses its CO_2 to the surrounding air—this means that older coffee will be easier to extract evenly and will produce espresso with less crema. This loss of CO_2 can be useful to determine coffee freshness in a quantitative way (e.g., Smrke et al. 2018), but it is also useful in practice because it can push oxygen away from a coffee bag and form a protective CO_2 environment around the beans, assuming they are stored in a bag that has a one-way valve. Unlike oxygen, CO_2 does not react chemically with the content of the coffee beans, which is why it protects the coffee and why we call it an "inert gas."

8.2 Aromas

Like CO_2, several chemical compounds responsible for coffee aromas are volatile and will degas as the beans age. Volatile compounds are initially retained in the coffee oils or adsorbed[42] on the internal cellular walls of the coffee bean, and they will degas to the surrounding environment at a rate that depends on the temperature and the pressure difference between the inside of the coffee bean and the environment outside it. The presence of moisture will also have a significant effect on how fast aromas are liberated (Anese, Manzocco, and Nicolli 2006).

Because aromas degas at different rates, the smell and taste of coffee will evolve over time as it slowly stales (e.g., see Marin et al. 2008). These changes in aromatic profile will not necessarily be negative in the short term, but they may become displeasing over the longer term and will be accompanied by a reduction in intensity of the coffee aromas, unsurprisingly. Because of this, different coffee beans prepared with different methods may taste best at different moments as they age (e.g., Smrke 2020). It is interesting to note that darker-roasted coffee will tend to lose its aromas faster because of its larger internal pores (Schenker et al. 2000).

8.3 Coffee Oils

Coffee oils are initially located inside the coffee beans and will slowly migrate to their surface because of capillary action, causing the *oil sweating* effect. We discussed the phenomenon of capillary action in Chapter 4 in the context of water reaching inside coffee particles during the blooming phase. This same phenomenon causes coffee oils to migrate outward to the bean's surface and will be aided by the internal pressure that was initially stored inside the beans during roasting (Folmer 2017). Because oil is much more viscous than water and the path to the surface of a whole coffee bean is longer than the path to the surface of a ground coffee particle, this phenomenon will take not minutes, but rather days or even months, depending on the roast degree. Darker roasts and decaffeinated coffee will tend to show this effect faster. Exposed coffee oils

[42] retained on the walls of a solid as a thin film

can oxidize by reacting with oxygen in the ambient air (e.g., Smrke et al. 2018), causing an unpleasing rancid taste. As coffee oils begin appearing on the surface of a coffee bean, they will become susceptible to this effect of oxidation. Even though fresh coffee beans usually do not seem oily, it can take only about a day for microscopic droplets invisible to the human eye to appear on the beans' surface (Schenker et al. 2000).

8.4 The Enemies of Coffee Freshness

Now that we have discussed some of the ways in which coffee can stale, we can begin identifying what factors can worsen it or slow it down. The five main enemies of coffee freshness are

- Time
- Heat
- Moisture
- Oxygen
- UV light

Time is the most obvious factor; if you wait long enough, the majority of aromas will escape, oil will sweat out and oxidize, and the coffee will end up tasting stale. In many regards, heat is similar to time because heating up the beans will speed up degassing and chemical reactions and will make oil less viscous, favoring its escape to the surface. Both moisture and oxygen may cause damaging chemical reactions to the chemicals inside the coffee beans, and moisture, in particular, will also favor the loss of aromas (Anese, Manzocco, and Nicolli 2006; Cardelli and Labuza 2000). However, some interesting preliminary experiments show that controlled exposure to moisture can affect the taste of roasted coffee in a subjectively pleasing way (Hoffmann 2020, Jan. 20; Apr. 28). Long-term exposure to UV light can also favor chemical reactions detrimental to the stability of the coffee and other food products, for example, by damaging proteins or favoring the oxidation of oils (Csapó et al. 2019).

8.5 Preservation Strategies

Now that we have identified the factors that can damage coffee freshness, we can devise strategies to preserve coffee longer at its optimum flavor profile. The most obvious preservation method is often the package that the coffee beans come in. Many roasters take care to select bags made of airtight materials and use a one-way valve to avoid damage to the bag that could follow a buildup in pressure as the beans degas. Some roasters will even flush their bags with inert gases to push moisture and oxygen outside completely before sealing the bag. In the presence of a one-way valve, the natural degassing of CO_2 by coffee will achieve a similar result if the coffee was placed in the bag shortly after it was roasted. One-way valves are not an ideal solution because they can also let aromas out. In principle, beans flushed with inert gases in a sealed bag will probably remain fresh longer—until the seal is broken when the bag is first opened, at which point the beans will be more vulnerable to moisture and oxygen because CO_2 will not be able to push them out.

In situations where coffee is left in a good-quality bag without ever breaking the initial seal, it can remain fresh for a surprisingly long period of time. For example, I was not able to distinguish a fresh bag from a 3-month-old bag of the same bean and roaster in a blind tasting. This result was initially surprising to me, but I later realized that we are accustomed to thinking that roasted coffee "goes bad" quickly because it does so when a bag is opened many times. Essentially, a good-quality bag is a great device for preserving coffee as long as it remains sealed, because the initial CO_2 degassing acted like an inert gas flushing of the coffee bag. Every time a coffee bag is opened, new moisture and oxygen are introduced back in, at moments where the coffee has little CO_2 left to push them out.

Inert gas canisters are often available at specialty wine stores, and they can be used to flush a bag free of oxygen and moisture just before closing it, slowing down the aging process. This can, however, get wasteful and expensive very fast, so I would only recommend it for longer-term preservation of expensive coffee beans.

Another factor that makes the preservation worse is the growing headspace in the coffee bag every time some of the coffee is consumed. A larger headspace means more oxygen and moisture can potentially damage each coffee bean and that there is more available space for CO_2 and aromas to degas before the process is slowed down by an increase in the environment pressure in the absence of a one-way valve. Bags are generally preferable to rigid containers for this reason, as air can be pushed out to remove some of the headspace.

Based on these considerations, some manufacturers have designed options for storing coffee either in a canister with a piston-like movable lid to remove the headspace (e.g., the Airscape canister), which achieves similar results to a bag in which headspace has been carefully compacted. Others, such as the Fellow Atmos canister, go a step further by adding a mechanical valve to create a weak vacuum and reduce the oxygen and moisture content of the headspace.

This solution will likely slow down oxidation and moisture damage, but it may speed up the degassing process, especially when the headspace becomes large. A simpler yet elegant solution is to single-dose coffee beans immediately, when a roaster bag is first opened, into small containers with a small headspace. This allows each dose of coffee to have withstood only a single event of exposure to air and moisture. Weber Workshops designed such single-dosing cellars that have a one-way valve, further allowing the single-dosed beans to push away any oxygen and moisture with the remaining CO_2 they will degas, assuming they are fresh enough when they are transferred to the cellars.

In more unusual scenarios, coffee can be stored with inert gases under pressure, which can significantly slow down the degassing of CO_2 and aromas (e.g., see Illy and Viani 2004). In principle, this could be one of the most efficient preservation strategies, especially when combined with freezing, but it is much harder to achieve in practice because it requires the use of pumps or pressurized gas canisters as well as inert gases. It is an expensive and wasteful solution, and it becomes important to use robust materials to avoid breakage of the canisters.

Vacuum-sealing coffee can be a great solution to extend its shelf life by many months, but it is more wasteful, and I would only recommend doing so in soft rather than rigid packaging. Otherwise, creating a vacuum can compound the negative effect of the headspace and cause the coffee aromas to be sucked out from the beans over time. When beans are stored under vacuum in a soft bag, the bag will press against the beans because of the outside atmospheric pressure pushing down on the bag. This will slow down the degassing of the coffee beans and ensure they do not encounter moisture or oxygen.

However, the process of vacuum-sealing probably sucks out some aromas in the short time required to create the vacuum, and for that reason it would be preferable to avoid vacuum-sealing coffee beans multiple times. More expensive chamber vacuum-sealers can also be used to reach much stronger vacuums than the cheaper Food Saver® devices, further extending the shelf life of coffee. I used a relatively cheap vacuum device for years, and in my experience, it really helped to preserve coffee for long periods of time, but it did cause a sometimes-noticeable degradation in aromas when the coffee was vacuum-sealed. I noticed this effect the most with Ethiopian coffees, perhaps because they have more noticeable volatile aromas.

Freezing is one of the most obvious and powerful methods to preserve coffee freshness for a long time. When done properly, it can preserve coffee freshness for up to a year, or even more. One must be very careful when freezing coffee, however, because it can easily expose coffee beans to significant moisture. Every time a freezer door is opened, moist air can make its way in, and that moisture will rapidly condensate on the surfaces in the freezer because of the cold temperature. This can cause a layer of ice crystals to form on anything that is not properly sealed, including most plastic containers and Ziploc® bags. One-way valves on coffee bags are also not designed to function in cold environments because they rely in part on a thin compartment of oil, the properties of which can be significantly changed at cold temperatures. Other types of one-way valves, such as the silicon dome valves on the Weber Workshops cellars, can work at freezer temperatures.

When coffee beans are removed from a freezer, they can remain cold for hours, depending on how they were packaged. If the beans are then exposed to air while they are still cool, some moisture will immediately condensate on their surface. This is not a problem if the coffee is immediately ground and brewed, but it can lead to rapid degradation if the coffee is then left in a bag with this additional humidity. These considerations mean that freezing coffee should be done carefully; beans should always be sealed completely, and only coffee to be used immediately should be exposed to air while still cold. Freezing likely does not damage the internal structure of coffee beans, and I am not aware of any data suggesting that multiple freezing cycles can damage coffee.

As we discussed in Chapter 3, grinding coffee beans while they are still cold is thought to have beneficial effects on taste (e.g., Uman et al. 2016). It is likely that doing so produces slightly more coffee fines (below 100 μm in size). Fines may be detrimental to the quality of percolation, but in my experience, grinding frozen beans still seems to improve the perceived quality of filter coffee, likely caused by a better preservation of coffee aromas during grinding.

We have already mentioned that heat is an enemy of coffee beans, but it is worth repeating how important it is to keep any coffee container away from heat at all times. Labuza et al. (2001) calculated that the shelf life of coffee can drop from about 21 weeks at 22°C (72°F) down to 16 weeks if it is stored at 35°C (95°F) in the same package (see also Illy and Viani 2004).

Oxygen or moisture scavengers are often used to protect dry food from the environment. Oxygen scavengers come as small packets that contain powdered iron and other compounds that accelerate the capture of oxygen. I used them for a few months before stopping completely when I realized they began to affect the flavor of coffee, especially when they were kept at room temperature. Yam (2009) noted that some oxygen scavenger packages include moisturizing agents to help the capture of oxygen by the powdered iron, and they may even absorb CO_2 preferentially instead of oxygen when placed in CO_2-rich environments such as coffee bags. It is possible that reactions with coffee aromas may even cause a cascade of secondary reactions that will negatively affect the taste of brewed coffee. Moisture scavengers are usually made of silica balls that can adsorb moisture inside a porous structure with a vast internal surface, but they are also likely problematic if they adsorb coffee aromas and liberate the headspace for further degassing.

At this point, you may be wondering what the best way is to implement all of these strategies to preserve coffee freshness. This will depend on your habits, the price of coffee you are drinking, how long you plan to store it, and many more factors. I would therefore encourage you to consider a customized solution for your needs based on the knowledge presented in this chapter. The following recommendation may be generally good for home users: if you receive a coffee order containing three or more different bags, for example, resist the temptation to open them all at once. If you can drink them in sequential order, each bag will remain much fresher. Also ensure that the bags are stored away from heat, light, and moisture.

I like to store my coffee beans in a low and dark kitchen cabinet located far from the oven and any heat-generating compressors. I often leave the beans in their original roaster bag for a week or so before opening it. I ensure my hands are dry when I open the bag, and I spoon the beans out of it, as CO_2 is slightly heavier than air and this will favor protecting the beans from oxygen and moisture. I always push out all air from the coffee bag when I close it. When I plan to keep coffee for longer than a few weeks after the roaster bag was first opened, I split them across small bags or Weber cellars as soon as I first open the roaster bag. If I plan to keep some coffee for several months, I usually vacuum-seal it either as a single dose or in larger bags and store them in the freezer. I also always keep a few high-quality empty bags with a robust Ziploc-like mechanism and a good overall seal except for the one-way valve, like those used by the more careful roasters, and I use them to store any coffee that comes in a low-quality container such as a paper bag.

CHAPTER 9

Roast, Terroir, Varieties, and Processing

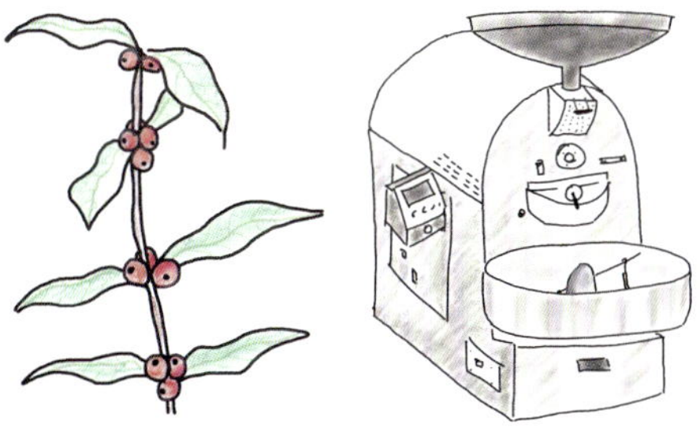

So far in this book, our discussion of coffee brewing has assumed that all coffee beans behave the same as we grind and brew them. This is far from the truth; coffees from different origins and varieties have different properties and different taste profiles, and they will produce different average extraction yields. This chapter gives a summary of these differences in bean properties and explains how baristas can adapt their brewing accordingly. We will not dig very deep into the science of coffee varieties, processing, and roasting. For more detailed discussions of these subjects, I recommend the following books: Bekele and Hill (2018), Brown (2018), Hoffmann (2014), Rao (2014), and Rao (2020).

This chapter will review some practical effects that different coffee varieties, origin countries, roasting level, and methods of processing have on coffee brewing. These observations are based on compilations of coffee bags that were recently selected by green coffee buyers among specialty coffee roasters and are thus biased toward current trending preferences. As roasting style, detailed green coffee processing methods, farming practices, and roasters and consumer preferences tend to change over time, our observations may not hold true forever, depending on their exact causes.

Furthermore, all the aforementioned factors are intertwined in a complicated way—for example, different origin countries have different farming and processing practices. Therefore, the correlations we will discuss will generally not allow us to determine which exact aspect of a coffee bean caused the observed consequences on the cup of coffee we brew. Rather, they will provide guidelines for consumers of specialty coffee like me, applicable now and perhaps for not longer than a decade. If we determine that some coffee varieties result in higher average extraction yields or generate more fines when grinding, not knowing whether the true cause was a different farming practice or a genetic difference will not prevent us from adapting our coffee brewing in a useful way, for as long the farming and roasting landscape has not changed in a significant way.

9.1 Grinding, Hydraulic Resistance, and Clogging

In Chapter 3, we extensively discussed how different grinders generate different particle size distributions, which in turn directly affect the taste profile and the flow of a pourover brew. We also learned in Chapter 4 that the presence of very fine coffee particles can lead to partial clogging of the bottom parts of the coffee bed or the paper filter, resulting in uneven extractions that can taste harsh or astringent. We briefly discussed that the coffee bean properties can also affect the resulting particle size distribution, although these effects are typically smaller than changing grinder or burr geometry.

The bean properties relevant to grinding include their temperature, their brittleness (i.e., how easy it is to break them), their mass density, their hardness, and their moisture content. All these quantities can change how easily the beans can be cut or shattered and may also affect how much heat is transferred to them during grinding. Although it is hard to measure these effects, we can characterize the way different beans or different roasts react to the same grinding parameters. To investigate this, I ground several different coffee beans directly from the freezer, at the same setting on the EG-1 v2 grinder with SSP ultra-low-fines burrs.[43]

As the figure opposite shows, the target peak location was approximately the same for all beans, but its exact shape and narrowness showed variations not entirely accounted for by uncertainties that represent my limited sample sizes. The quantity of boulders varied very slightly, and the quantity of coffee fines varied somewhat more. It is interesting to note how the values of D_{10}, the parameter that correlates with the flow rate of water through the coffee bed, changed quite significantly. This does not necessarily mean that grind sizes should have been adjusted to align all D_{10} values, because my analysis cannot characterize fines smaller than 100 μm, and changing grind size affects the taste profile in a way that is partly independent of the flow rate of water. In other words, different coffee

[43] The zero-point of my grinder has changed in this figure, compared with other figures in Chapter 3: grind size 4.7 here corresponds to a burr spacing of 320 μm, similar to grind setting 7.0 in Chapter 3.

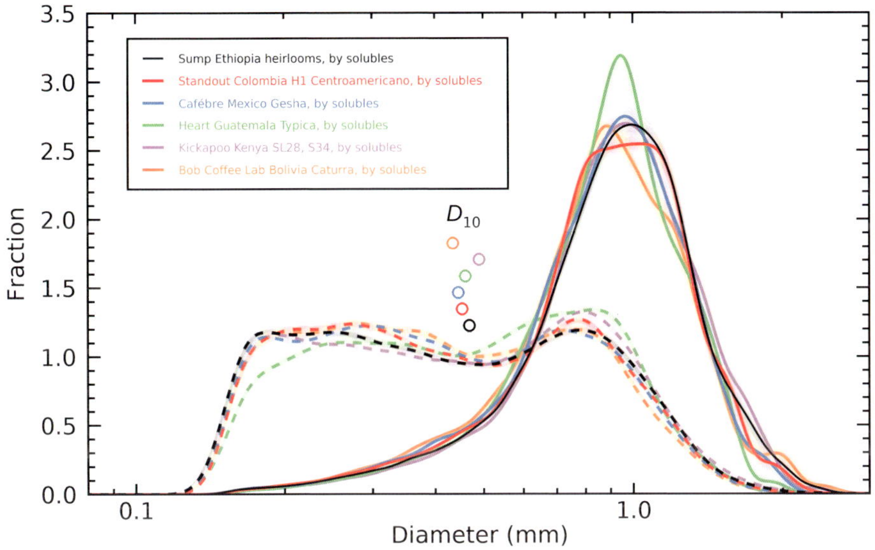

Particle size distributions obtained with the EG-1 v2 with SSP ultra-low-fines burrs for various frozen coffee beans. Dashed lines show the number of particles of each size, and full lines show their relative contributions to the total solubles. The values D_{10} indicated with circles correlate with the coffee bed permeability produced by these specific particle size distributions.

beans may taste optimal with different brew times, but obtaining similar brew times would require adapting the grind size, as is widely done in the context of espresso preparation. I have found that very few coffees taste better when I grind finer than the grind size shown above, but this could be a matter of personal preference and could also depend on the specific grinder and burrs that I use.

Inspecting a carefully selected subsample of 416 V60 brews in my coffee log—each prepared in exactly the same way and with the same grind setting, but with 192 distinct bags of coffee—indeed shows that different origins tend to produce different brew times on average.

This may be explained by different origins generating different particle size distributions on average. Ethiopian coffees seem to be the most prone to produce a significant amount of fines that result in longer brew times and brews more susceptible to clogging. This effect may come from the fact that Ethiopian green coffees have a low moisture, which likely leads to drier and more brittle roasted beans. Both Ethiopian and Kenyan green coffee beans tend to be much harder than the rest, as I found from measuring the surface hardness of seven types of green coffee beans with a Shore D durometer.

These limited measurements suggest that bean hardness may not be the culprit for the longer brew times produced by Ethiopian coffees, because Kenyans do not also produce unusually long brew times, in general. The smaller

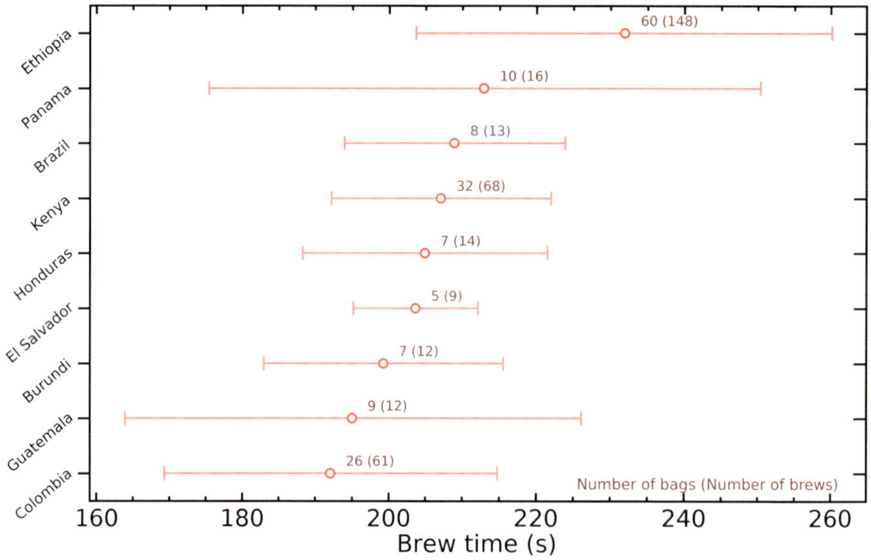

Average and standard deviations in brew time for coffees grouped by origin. Slight correlations are observed, with the most notable outlier being Ethiopian coffees that infamously often cause longer brew times and clogging.

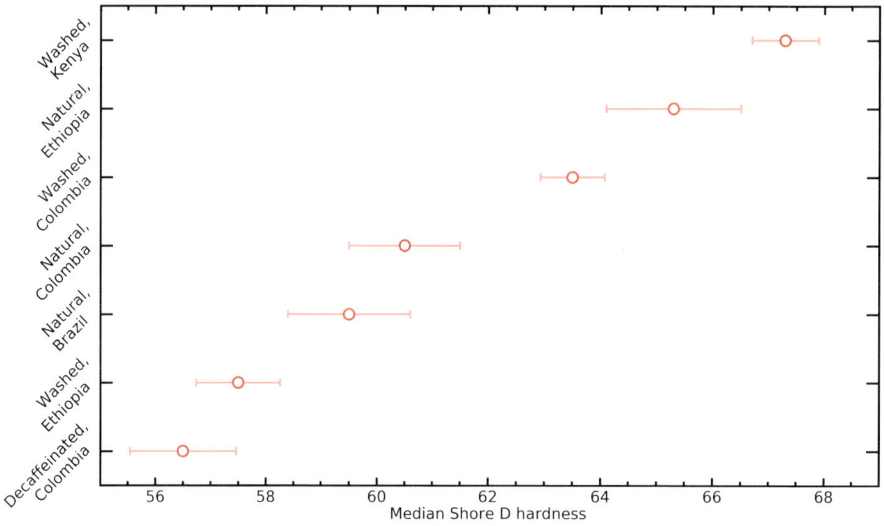

Average hardness of different green coffee beans by origin. Larger values correspond to a harder bean surface.

average bean size of Ethiopian coffees may also be a viable hypothesis to explain this observation.

Another important factor to consider is the roast profile. To demonstrate this, I measured the particle size distributions of two coffees roasted differently:

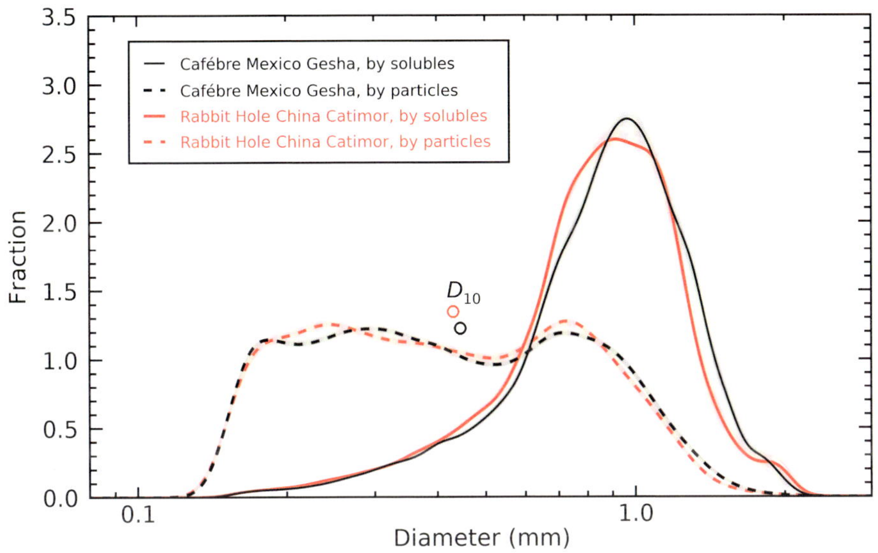

Particle size distributions of a light-roasted Mexican Gesha compared with a darker-roasted Chinese Catimor

a light-roasted Mexican Gesha by Cafébre and a medium-roasted Chinese Catimor by Rabbit Hole Roasters.

As could be expected from our discussion of coffee brittleness, it seems that the darker-roasted Chinese Catimor produced more fines and a slightly finer target peak size. To illustrate the difference in roast styles, I brewed the Mexican Gesha at a kettle temperature of 99°C (210°F) and the Chinese Catimor at 90°C (194°F), using a lower kettle temperature to avoid bitterness in the latter case.

Inspecting my brew logs for the 133 coffee bags for which I measured bean color with a reflectometer (as described in Chapter 11) does not reveal longer brew times for darker roasts; it may actually show a reverse correlation for which I do not have a definite explanation. It might be caused by less-hard coffee beans such as those from South America often being roasted darker, for example.

In the figure at the top of the following page, a greater roast development is indicative of darker beans that reflect less light, as described in more detail in Chapter 11. A supermarket French roast would have a roast development value of about 85%, whereas a Nordic light roast would be in the range 65–70%.

Decaffeinated coffee tends to be extremely brittle and produce a significantly increased amount of fines, leading to longer brew times and potentially clogging brews unless the grind size is adjusted to be coarser than usual. Inspecting my log of coffee brews by processing method shows that decaffeinated coffee produced much longer brew times than both washed and natural-processed

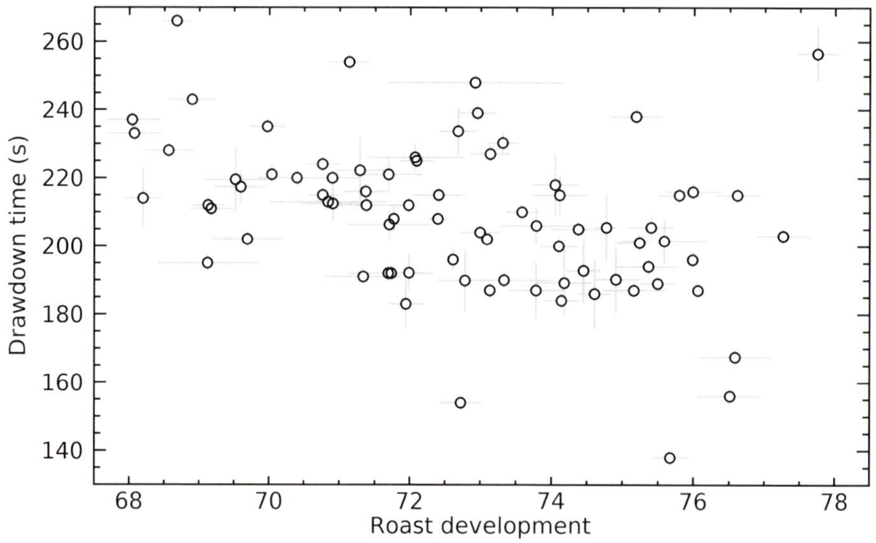

Average drawdown time as a function of roast development measured from bean color, grouped by individual coffee bags. On the horizontal axis, darker-roasted beans are farther to the right, and vice-versa. This figure shows that darker roasts tend to produce faster brew times.

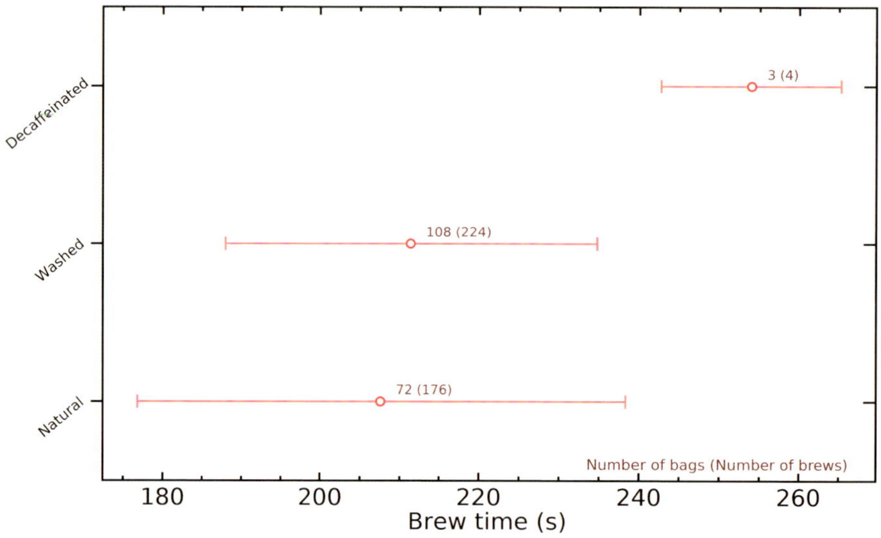

Average and standard deviations of brew times, grouped by coffee processing method. No large difference is observed between washed and natural-processed coffees, but decaffeinated coffees produce notably longer brew times for the same grinder setting.

coffees, while the latter two did not differ significantly from each other.

An example particle size distribution is shown opposite for regular and decaffeinated Kenyan coffees; the decaffeinated coffee indeed generated more fines, as well as more boulders, for a smaller D_{10} value (described in Chapter 3), and thus a slower predicted drip rate.

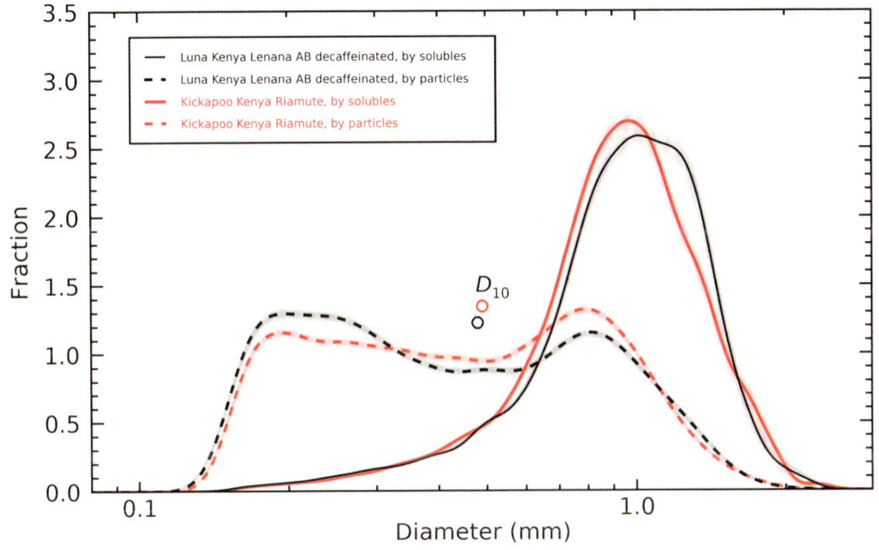

Particle size distributions for regular and decaffeinated Kenyan coffees, ground with the same grinder setting. The decaffeinated coffee generated both more fines and more boulders, consistent with the observation that it drips more slowly.

Hoffmann (2020, Apr. 28) noted that coffee beans aged in a high-humidity environment produced faster drawdowns when brewed as V60. This may be yet another instance of more-ductile coffee generating less fines during grinding.

9.2 Available Solubles

If you habitually measure your coffee concentration with a refractometer and calculate the average extraction yield of every coffee you drink, you will eventually notice that different beans produce very different average extraction yields when treated the same way, even when the brew time does not change significantly. This could be explained in part by small changes in the particle size distribution that do not affect the average coffee bed resistance, but I believe the biggest factor to explain this is more intrinsic to the coffee beans themselves.

I currently have two leading hypotheses to explain this observation that could be tested with dehydration experiments to directly measure the total dissolved solids (TDS) rather than deducing it indirectly with a refractometer. The first possibility, which I believe is the more likely one, is that different coffee beans simply have different fractions of soluble chemicals. The second possibility is that different chemical compositions of specific coffee types change how the refractive index of coffee relates to its total dissolved solids, which would mean that some specific coffees bias our estimation of the true total dissolved solids in different ways, and would not necessarily mean that some coffees pro-

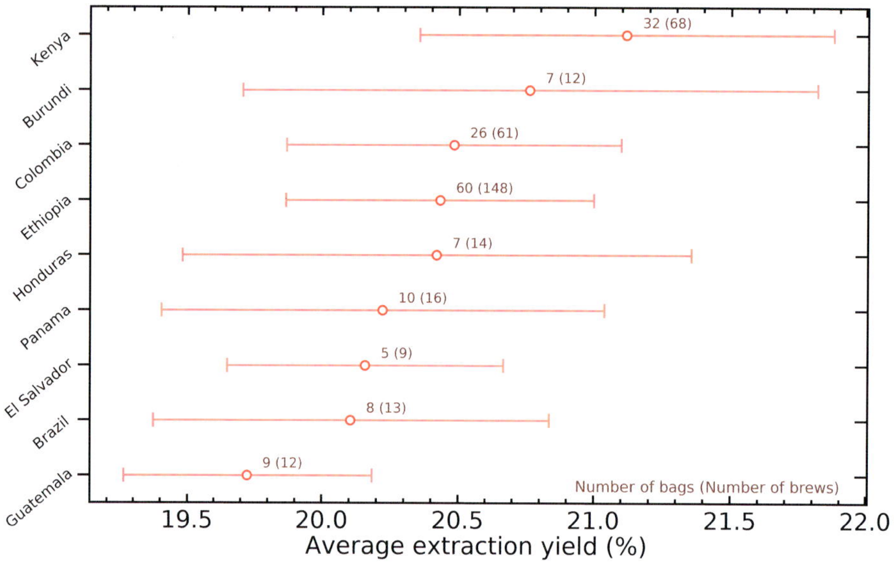

Average value and representative ranges (standard deviations) of average extraction yields obtained with different coffee beans grouped by origin and prepared identically as V60

duce brews that are truly more concentrated. Inspecting the average extraction yields that I logged for my V60 brews demonstrates these variations quite clearly.

As the figure above demonstrates, origins such as Kenya, Burundi, and Africa, in general, seem to produce brews with systematically higher average extraction yields, compared with coffee from Central and South American countries. I recently had an experience that illustrated this quite starkly, where a 1:17-ratio pourover of a Kenyan SL28 and SL34 Waturi AB roasted by Lüna produced a 1.70% TDS coffee concentration, immediately followed by an Anaerobic fermented Catuaí from Los Lajones in Panama roasted by Standout Coffee Roasters that produced a 1.36% TDS cup. Both coffees were delicious, well roasted, and prepared with the same brew method.

Inspecting the data across the dimension of growing altitude, shown opposite, is insightful. I used larger circle sizes for points located in denser regions of the figure, to move visual focus away from single outliers, and blue curves represent the 1, 2 and 3σ covariances of the data.[44] One underlying mechanism that could explain this 25.6% Pearson correlation coefficient between elevation and average extraction yield is based on the observation that green coffee cherries tend to con-

[44] Those curves are ellipses oriented along the correlation and with aspect ratios that capture the spread in both directions. The 1, 2, and 3σ contours include, respectively, 39%, 86%, and 98.9% of the data when it is normally distributed. Some more-informed readers may be surprised at these exact fractions; the usual are 68%, 95% and 99.7% when working in one dimension.

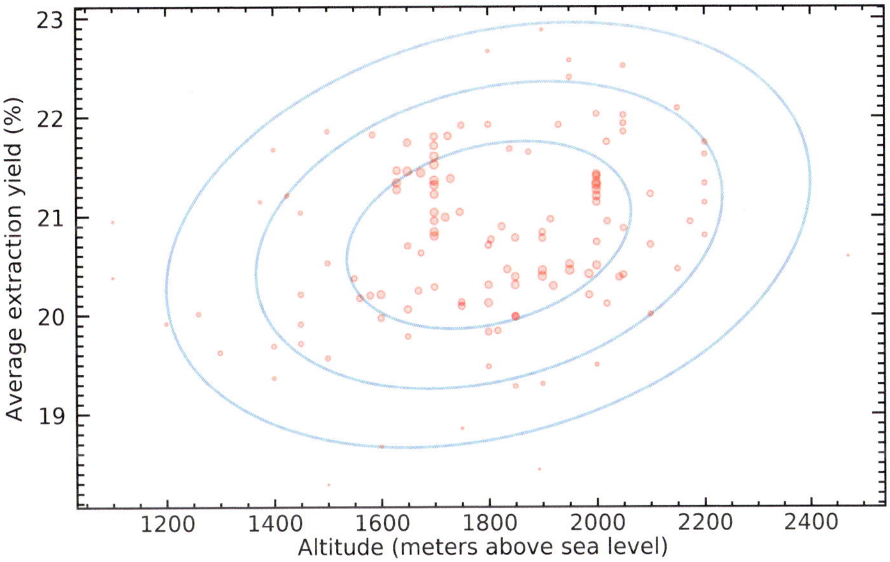

Correlation between the average extraction yield and growing altitude of coffee beans

centrate more nutrients in their seeds in cooler, higher-elevation environments (Vaast et al. 2006; Worku et al. 2018). If this mechanism is indeed responsible for the largest part of the average extraction yield differences between different coffee beans, it would corroborate my first hypothesis: that there really are more chemicals to be dissolved in some beans compared with others, rather than this being an issue of chemical composition throwing off the refractometer calibrations.

It is also interesting to look at how average extraction yields are affected by the specific coffee variety (top of following page), as listed on the coffee bag.[45]

Inspecting the average extraction yield versus roast development for my past coffee brews (bottom of following page) reveals another interesting trend. This figure shows a clear correlation: contrary to popular belief, darker roasts seem to yield lower average extraction yields, a trend that seems to hold within each coffee origin. As you may have realized by now, this could be explained by the fact that darker roasts have shorter brew times, on average. Brew times seem to account for only a small fraction of this effect, however; even if we display only the brews with brew times between 3:00 and 3:30, we still see a clear correlation. I also measured average extraction yields and roast colors for six batches of the same green coffee roasted in different styles by Andy Kyres.

Yet again, a trend was observed, where the darkest roasts seemed to produce lower average extraction yields. Interestingly, in this case the lightest roasts

---

[45] Coffee bags with more than one listed variety were counted in each category.

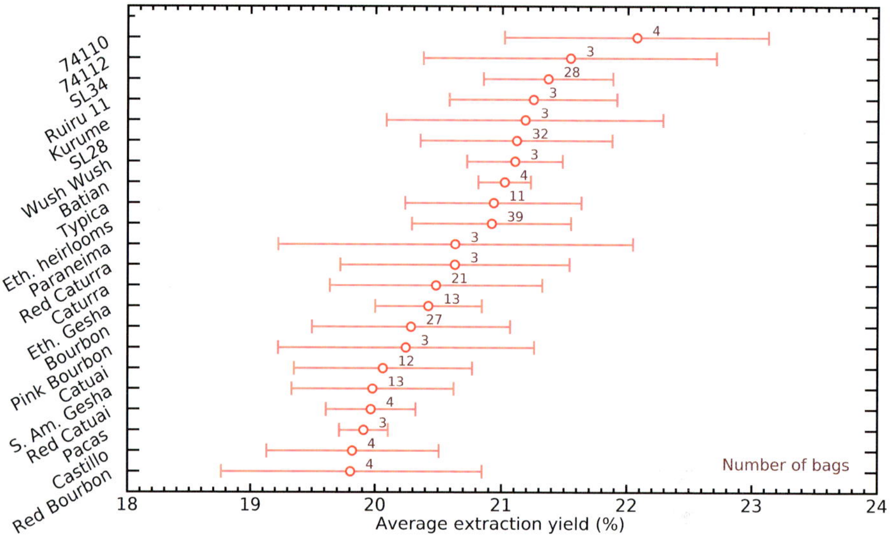

Average values and representative ranges (standard deviations) of average extraction yields obtained with different coffee beans grouped by variety and prepared identically as V60

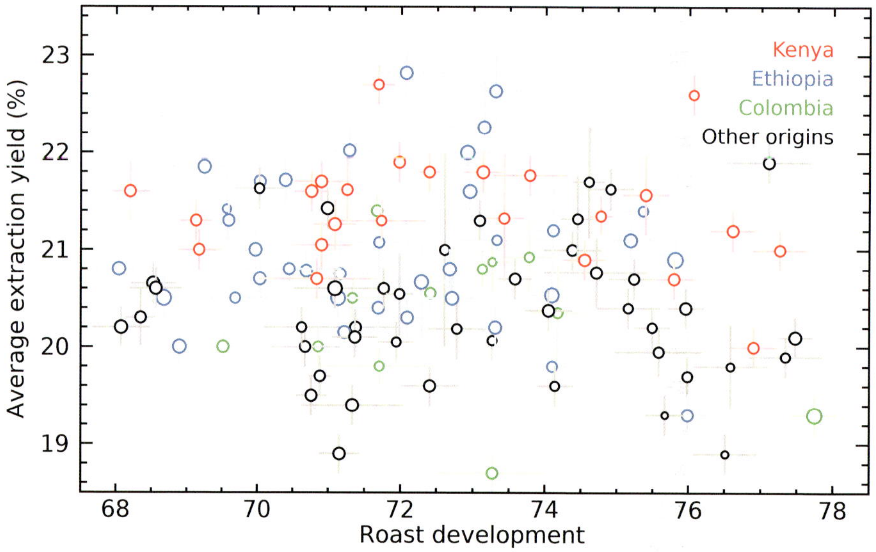

Average values and standard deviations of average extraction yields as a function of roast development, measured by bean color. Every circle represents a single coffee bag. Symbol colors indicate origins and larger symbols indicate longer average brew times. The data show a correlation: darker-roasted coffees result in lower average extraction yields independent of origin and brew time.

also seemed to result in slightly lower average extraction yields. It is possible that this trend may be different in the context of espresso, where brew times are much shorter. Darker roasts that enlarge pores between coffee cells may prove to be crucial to accelerate diffusion inside coffee particles and rapidly bring their

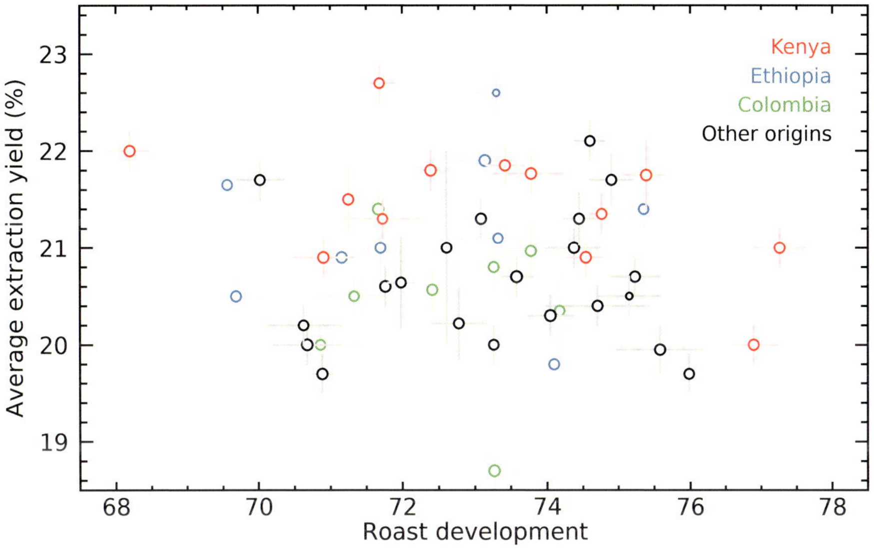

Average values and standard deviations of average extraction yields as a function of roast development, measured by bean color. Every circle represents a single coffee bag. Symbol colors indicate origins and larger symbols indicate longer average brew times. Only brews with brew times between 3:00 and 3:30 were considered here. The data still show a correlation: darker-roasted coffees result in lower average extraction yields.

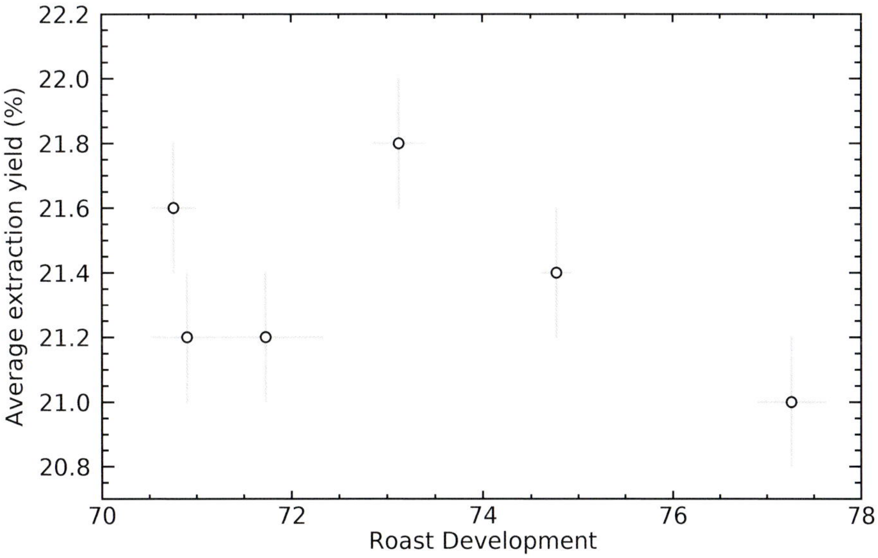

Average extraction yields measured for roasting batches of different roasting development using the same green Kii AA coffee from Kenya , roasted by Andy Kyres

solubles to the surfaces of coffee particles for classical espresso preparation to reach adequate extraction yields. Repeating this experiment with different brew methods may therefore yield different results.

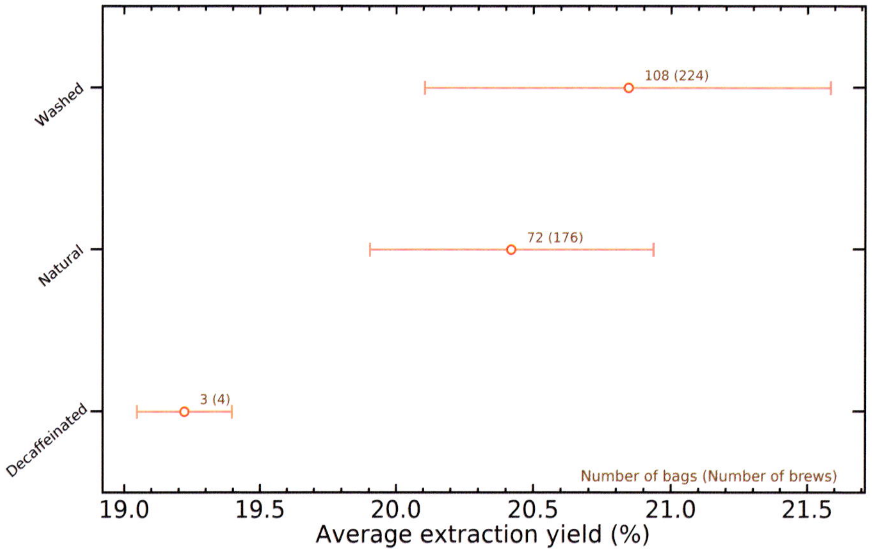

Average values and standard deviations for average extraction yields of brews, grouped by processing method. Natural coffees seem to extract slightly lower on average, and decaffeinated coffees show significantly lower average extraction yields.

It is worthwhile to note that decaffeinated coffees have the least available solubles, by far. This is not especially surprising, given that the coffee has already been depleted of a part of its chemical content (ideally, mostly caffeine) during the decaffeination process. This effect can be clearly seen in my log of coffee brews, as shown above.

9.3 Taste Profile

One of the most notable and important differences between coffee varieties, origins, and processing methods is their distinct taste profiles. Some specific coffee varieties, such as the Kenyan SL28 and Ethiopian Gesha, are renowned in the coffee community for their distinct flavors—in this example, for their redcurrant and berry or their peach-like and floral characteristics, respectively. Analyzing user-submitted and roaster-submitted tasting profiles for 1500 coffee bags gathered by the Firstbloom app and grouping them among custom taste categories, I built some distinct representations for the coffee flavors of different coffee beans grouped by varieties or processing methods. In these visualizations, the area of each subcategory is inflated with respect to how often a taste characteristic came up in describing a coffee of this type. I have reproduced some of these "flavor wheels" on the following pages.

Flavor Wheels—by Coffee Variety

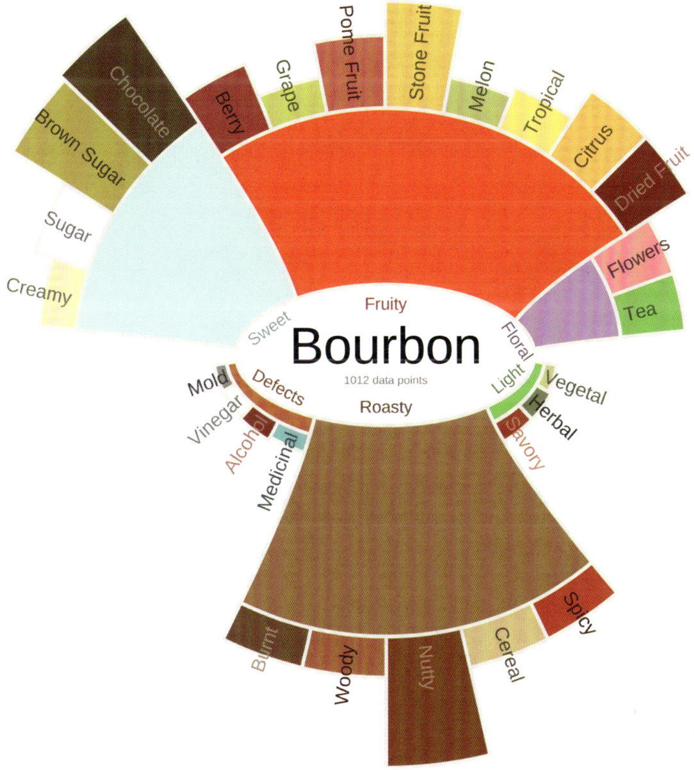

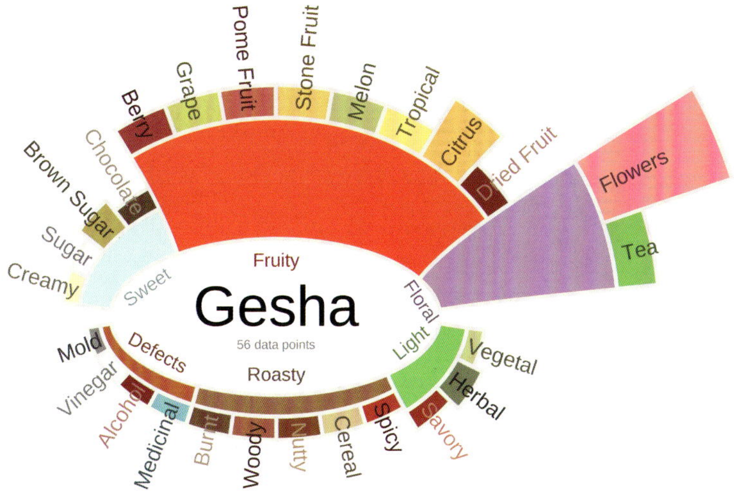

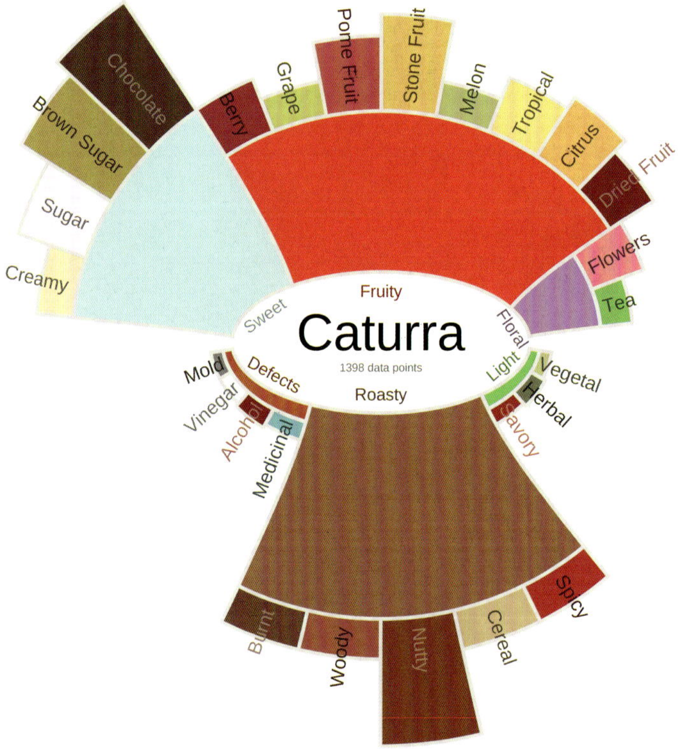
Caturra
1398 data points
Fruity
Sweet
Floral
Light
Roasty
Defects
Berry
Grape
Pome Fruit
Stone Fruit
Melon
Tropical
Citrus
Dried Fruit
Flowers
Tea
Vegetal
Herbal
Savory
Spicy
Cereal
Nutty
Woody
Burnt
Medicinal
Alcohol
Vinegar
Mold
Creamy
Sugar
Brown Sugar
Chocolate

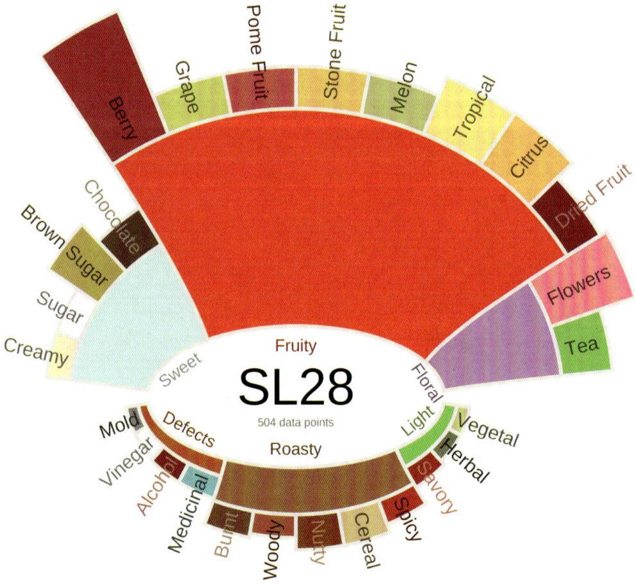
SL28
504 data points
Fruity
Sweet
Floral
Light
Roasty
Defects
Berry
Grape
Pome Fruit
Stone Fruit
Melon
Tropical
Citrus
Dried Fruit
Flowers
Tea
Vegetal
Herbal
Savory
Spicy
Cereal
Nutty
Woody
Burnt
Medicinal
Alcohol
Vinegar
Mold
Creamy
Sugar
Brown Sugar
Chocolate

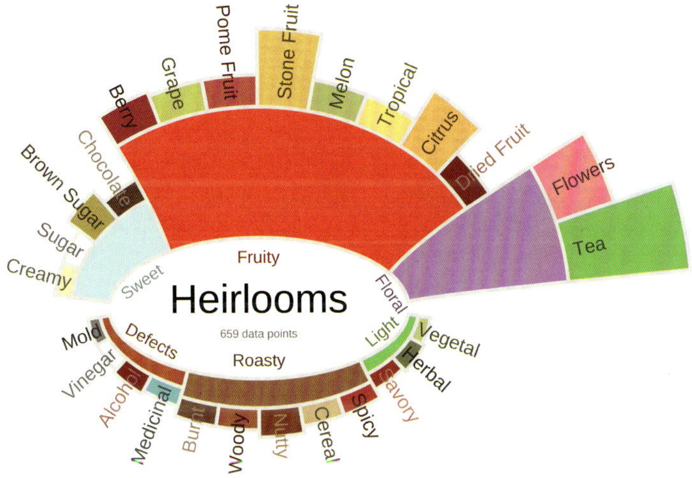
Heirlooms
659 data points
Fruity
Berry
Grape
Pome Fruit
Stone Fruit
Melon
Tropical
Citrus
Dried Fruit
Floral
Flowers
Tea
Sweet
Chocolate
Brown Sugar
Sugar
Creamy
Light
Vegetal
Herbal
Savory
Roasty
Spicy
Cereal
Nutty
Woody
Burnt
Defects
Mold
Vinegar
Alcohol
Medicinal

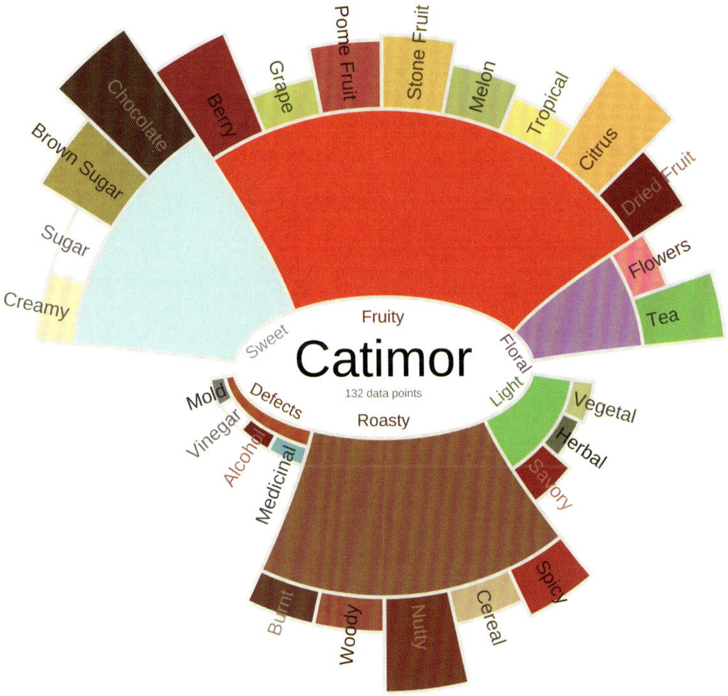
Catimor
132 data points
Fruity
Berry
Grape
Pome Fruit
Stone Fruit
Melon
Tropical
Citrus
Dried Fruit
Floral
Flowers
Tea
Sweet
Chocolate
Brown Sugar
Sugar
Creamy
Light
Vegetal
Herbal
Savory
Roasty
Spicy
Cereal
Nutty
Woody
Burnt
Defects
Mold
Vinegar
Alcohol
Medicinal

Flavor Wheels—by Processing Method

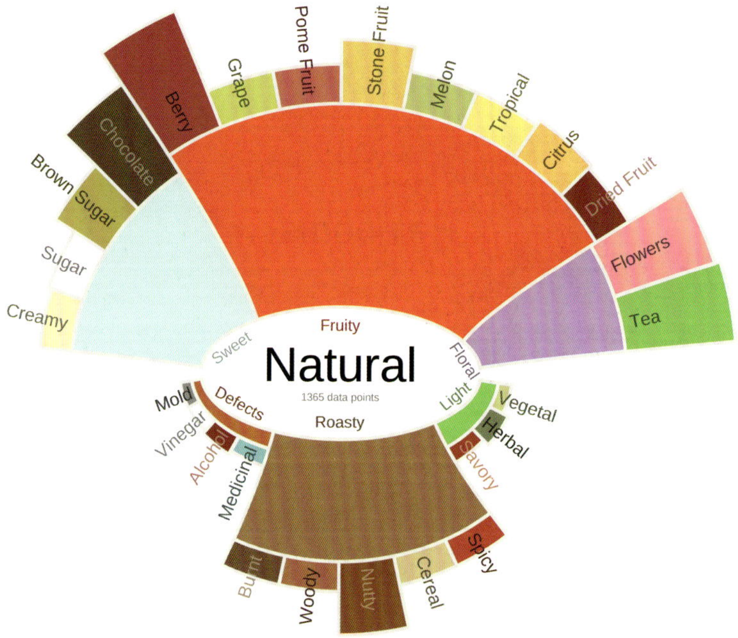

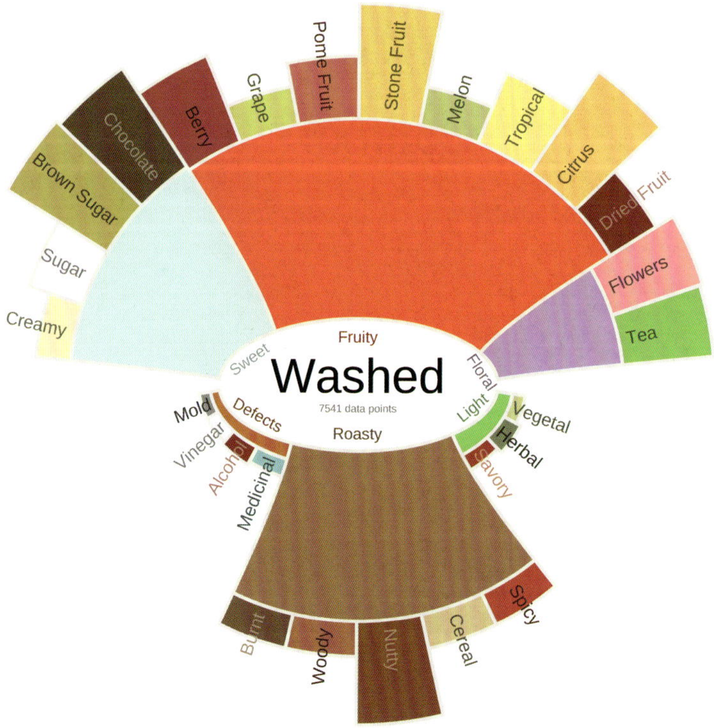

Another interesting way to represent the data is to choose a flavor descriptive and order the coffee types by how often they have been associated with that descriptive. See these examples that I found interesting below.

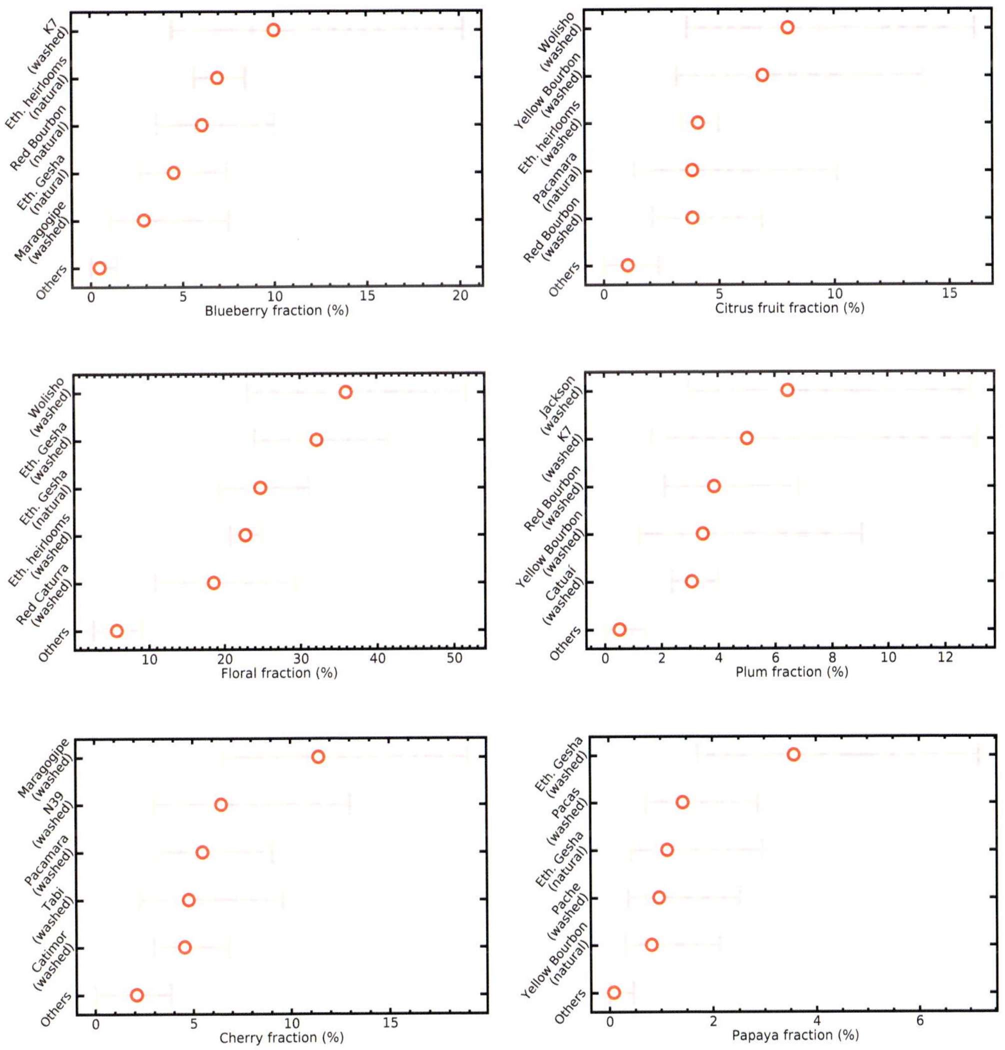

Fraction of coffee bags with an occurrence of specific taste descriptors. The horizontal error bars indicate uncertainties related to the limited sample sizes.

CHAPTER 10

Technique and Practical Applications

PHOTO BY NOÉ AUBIN-CADOT

Now that we have studied many aspects of filter coffee brewing, we can begin to put them into practice. Having a mental model of the physical processes happening as we brew coffee can give us a powerful way to decide, for example, what variable we want to change next when we dial-in coffee and how we might try to design a specific brew recipe.

In this chapter, I describe habits that I consider crucial to achieving good consistency in brewing, and I discuss how changing some variables may change the results of a brew. I also describe the V60 method I currently use as a reference point, although I do not think it is likely to be the optimum recipe—if such a thing even exists. It is likely that slightly different methods will be needed to yield best results, depending on taste preference, coffee beans, and roast degree, for example.

10.1 Consistency

Consistent preparation of pourover coffee is not an easy task. The first step toward achieving it is to be humble and not trust your unaided abilities. For example, using a scale to weigh the dry coffee dose and the water dispensed on the slurry is crucial if you hope to maintain any kind of consistency. I recommend against volumetric measurements, as they are much less precise. Achieving good consistency is a valuable tool, even if you are not actively crafting a custom recipe, because it will allow you to identify problems more easily as they arise. You can follow these recommendations if you are trying to get consistent results:

Water Properties

- Be mindful that any changes in your water composition can cause a change in your brew results. For example, tap water composition can sometimes change suddenly and will usually slowly drift over the months or seasons.
- If you use custom mineral water that you keep in a large bottle, shake it well immediately before use.
- Rinse your kettle from time to time, and descale it when you notice deposits forming at the bottom.
- Avoid letting the water boil for a dozen minutes or more before you begin brewing, because heavy evaporation and mineral precipitation could affect your water properties.

Grind Size Distribution

- Pay attention to room temperature: a change of several degrees can cause a slight shift in grind size due to an increase or decrease in the beans' ductility. Warmer temperatures will result in faster drawdowns. In most climates, this will usually be an issue only in winter if your room is unheated or in warm summer if you don't use an air conditioner. Coffee shops often get warmer as the day goes, even despite air conditioners. One way to counter temperature shifts completely is to grind coffee directly from the freezer, although this requires sealing and freezing beans as single doses in order to not waste the rest of the beans once the seal is broken, as water from the ambient air can condensate on the cold beans.
- Frequent grinding over a short period of time can cause a shift in your grinder temperature and similarly affect your particle size distribution. If you use a grinder with a motor sited close to the burrs or one that grinds at low rotation rates, this will tend to happen more easily.
- Clean your grinder burrs regularly. This can be tricky with some grinders, but doing so every few weeks will improve taste and ensure that your grind size distribution does not worsen over time.
- Starting the motor before you let the coffee through will result in a more uniform particle size distribution. Regardless of your choice in this matter, do not alternate between starting the motor before or after letting the coffee in.

- Be sure your grinder burrs were seasoned (see Chapter 3); otherwise, your grind size will slowly drift toward coarser particle sizes as you grind the first dozen pounds of coffee.

Flow

- Pay attention to how you place your paper filter in a dripper. For example, folding the seam of a V60 filter can have a large effect on how much water bypasses around the coffee bed. Folding it thoroughly by pressing the filter seam against a flat surface with your fingers can reduce bypass, and prerinsing the dry filter with tap water will make the filter stick more to the dripper walls, also reducing bypass. If you flip the dripper to pour water out before it drips, this will cause the filter to lift from the walls and will later increase bypass. Whatever method you choose to place and prerinse the filter, be sure you always do it the same way. The exact amount of water you use to rinse the filter is not important as long as you get rid of the paper taste, and always begin your brew with approximately the same dripper temperature.
- Level the counter where you brew coffee, or add shims so that your dripper is level without having to think about it every day. An off-level slurry will cause more water to flow on one side and make your extraction less even.
- Be aware that changing your paper filter brand or model can affect your brew results. One notorious example is the difference between the tabbed and tabless paper filters manufactured by Hario.

Slurry Temperature

- Use a kettle that displays and holds its temperature if you are not brewing off-boil. Try to have it reheating the water to your desired temperature between each pour.
- If you use a low-capacity kettle, consider adding an insulating coat to help it keep a more consistent water temperature.
- If you brew under a ventilation outlet, next to an air conditioner, or in windy conditions, your slurry temperature may be affected and lead to poor consistency.
- If you use a dripper with a high heat capacity (e.g., ceramic or glass), preheat it before brewing and always preheat it the same way—ideally, immediately before you put the coffee in and begin brewing. If you wait for too long between the preheating and brewing steps, you will start the brew at different temperatures, which will lead to different average extraction yields.

A ceramic evaporating basin used to sort out black beans and *quakers*

Execution

- Your brews will only be as repeatable as your coffee space is clean. Avoid allowing coffee grounds to accumulate in your grinder, and regularly clean and dry your equipment.
- Rinse and dry your dripper, coffee server, and coffee cups after each use, and wash them with a drop of fragrance-free soap from time to time. Be sure to use as little soap as possible, and rinse everything thoroughly. Avoid dishwashers, unless you can use them to disinfect your coffee equipment at high temperature without soap. It can be surprising how many rinses can be required after a coffee glass was washed in the dishwasher, even with fragrance-free soap. This effect can be worsened if your cups are made of porous material, such as ceramic.
- Try to always use the same server, countertop, and kettle. Changing the height of the dripper or the gooseneck diameter will make it harder to obtain consistent results.
- If you swirl the slurry during your brews, practice this skill: clog a filter with finely ground coffee, fill the dripper with water at various heights, and swirl until you are happy with the result. The slurry should move in circles rather than sloshing back and forth. You want to swirl the base of the dripper along small circles, and determine which rotation rate is appropriate, depending on the amount of water in the dripper.
- Be aware that switching from one bag of coffee to another can have import-

ant effects on your drawdown time and average extraction yield, even after you complete dialing-in the grind size. I do not recommend trying to obtain the same average extraction yield across different coffees or roasts.

- If you prefer to drink light roasts, you can use a plate or a ceramic evaporating basin to quickly inspect your coffee beans before grinding. Removing quakers (i.e., unusually pale coffee beans) or black beans can improve your consistency (as well as improve taste).
- Changing your dose by even small amounts (e.g., 0.5 g) will have a small but detectable effect on your brew, assuming you keep the same brew ratio. Changing the ratio will typically have a much larger effect on your brew, so keep your ratio as constant as you can.
- Try to always create a consistent coffee bed shape before brewing. I recommend using a nest shape for a more efficient bloom.
- Always bloom with the same amount of water and the same bloom time.
- Use a background visual aid so that you pour from the same height every time. One option could be placing a ruler on the wall so that you can always place the gooseneck tip at the same approximate height, from your perspective.
- Maintain a consistent pour rate. If you can afford to, purchase a scale that displays pour rate (e.g., the Acaia Pearl S). Otherwise, pay attention to the diameter of the kettle stream where it falls on the slurry, and try to keep it always the same.
- Decide on how many pours you want to use, and always use the same amount of water per pour from one brew to the next. Choose whether you want to initiate pours at a fixed time or at a fixed height of water in your dripper. I prefer the latter, as it allows me to maintain a water height of about 1 centimeter above the coffee bed for thermal stability. It is easier to be consistent with a recipe that requires less pours.
- Take a video of one of your brews and watch it again weeks or months later. You might be surprised to find how easy it is to slowly drift away from a brew method.

Measurements

- Choose a clear definition of brew time that allows you to measure it consistently. For example, I start the chronometer and begin pouring exactly at 5 seconds, and I stop the chronometer when the surface of the water passes below the coffee bed and I can see ambient light reflecting on the wet coffee grounds. I then subtract 5 seconds from the time I logged earlier and note this measurement as my brew time. In my experience, it is harder to obtain a consistent brew time than a consistent average extraction yield, because the former is very sensitive to exactly when and how you poured water.

- Be sure your coffee server is clean and dry before you brew. Using a wet or dirty server will cause you to measure your beverage weight inaccurately and may affect your average extraction yield calculation.
- Sample your beverage in a glass dropper (the sample will be used later for a measurement of total dissolved solids) immediately after you measure the beverage weight; otherwise, evaporation will affect your average extraction yield calculation.
- Wait for the dropper to cool for 15 minutes before you take a TDS measurement. Otherwise, you will affect the temperature of the refractometer prism, and you will obtain a biased measurement even if you wait for the temperature, and measurements, to settle.
- Letting your coffee sample cool in a glass dropper is better than letting it cool in a ceramic plate or another container, because the small area of coffee that is exposed to air in the dropper will ensure that very little evaporation takes place.
- Regularly clean your refractometer, and ensure its prism is clean and dry before taking a measurement.
- Always zero the refractometer moments before taking a coffee TDS measurement, using distilled water. Ambient temperature fluctuations will otherwise affect your reading. I like to keep a small distilled water dropper bottle next to my refractometer for this.
- Clean the refractometer prism with alcohol immediately after you take a measurement. I like to keep a small alcohol dropper bottle next to my refractometer for this.

10.2 Evenness of Extraction

One of the few goals I think we should strive for, regardless of taste preference, is an even extraction across the coffee bed. In general, this will tend to bring more clarity to the terroir flavors and increase the average extraction yield of a brew, assuming that nothing else has changed. One of the most powerful uses of a refractometer is to assess how a change in our brewing technique may have affected the evenness of extraction. In this case, we use the refractometer to look for increases in average extraction yield that cannot be explained by finer grounds, hotter water, or a longer contact time.

This is a subtler argument than saying a higher average extraction yield is always better. For example, I could use a 1:20 brew ratio, or I could grind extremely fine and achieve very a high extraction yield, with a resulting brew that tastes horrible, either because extraction was uneven or the flavor profile resulting from that combination of particle sizes is unpleasant. In my experience, higher extraction yields tend to taste better when they result from having extracted all parts of the coffee bed, leaving no underextracted pockets behind.

The details of what techniques may improve extraction evenness can be somewhat dependent on the specific grinder, coffee, and dripper that you are using. However, some general principles may help achieve an even extraction most of the time:

- **The bloom:** Performing a sufficiently long bloom step will help avoid dry pockets of coffee or excessive degassing during the later steps of the brew, as well as fully saturate the larger coffee particles with water. I normally bloom for 45 seconds, but the optimum duration of the bloom may depend on roast development; lighter roasts might benefit from longer blooms. Shaping the dry coffee bed into a "nest" by gently digging a central hollow with a finger will help make the blooming step more efficient.[46] Refer to Chapter 4 for technical details on the bloom. Note that blooming is not required if your brew begins with an immersion step (e.g., the AeroPress, Clever dripper, Hario SWITCH, French press, or siphon). In these scenarios, stirring or swirling the slurry as soon as water is poured on the dry coffee can still be useful to start the extraction evenly.
- **Swirling:** Gently swirling the slurry after every pour can help reset the channels forming in a bed of coffee and achieve a flat coffee bed surface that will favor a more even flow and extraction. Swirling can be more problematic and lead to clogging with grinders or coffees that generate more fines. Even in these scenarios, I would still recommend to swirl the bloom.
- **Coffee dose:** Choosing an appropriate dose for a specific dripper is important to achieve an even extraction. A too-shallow dose will favor the formation of channels, as described in Chapter 4; a too-tall dose may cause the bottom to be underextracted, and the additional amount of fines per surface area may clog the filter more easily. This latter point is discussed in more detail in Chapter 5. In Chapter 7, we learned how different drippers have different optimal

[46] This may seem surprising, but I found it was also true with flat-bottom drippers. I believe it allows prewetting of the bottom layers of the coffee dose more quickly, especially if one first pours water into the "nest" hole and then thoroughly swirls the dripper after enough water was added to the bloom.

doses, mainly because their geometry corresponds to different bed depths for the same coffee dose.

- **Kettle pours:** Pouring water in a way that favors turbulent agitation of the slurry will tend to help evenness of extraction, if it does not cause the brew to later clog. For more details, refer to Chapter 6. Avoid always pouring the water on the same location at the surface of the coffee bed. Doing so could cause a hollow, especially if you do not swirl the slurry, and this hollow will allow water to preferentially flow through that location, overextracting this region of the coffee bed and underextracting the rest.

Changing Brew Parameters

One of the most important skills for coffee brewing is being able to identify a problem and adapt your method to produce a good cup of coffee. This is a skill that can be worked on for years. For example, give baristas a completely new grinder that produces a drastically different particle size distribution, and you will quickly see how experienced they are at this.

Below, we review some of the parameters that are most often changed in preparing pourover. We consider how they may affect other brew variables as well as the resulting coffee:

- **Dosing:** The most important effect of increasing a coffee dose (with a fixed brew ratio) will be to make the coffee bed deeper. This will slow down the flow of water, as we saw in Chapter 4, but it will also increase the potential for clogging, as we discussed in chapters 4 and 5. If you want to brew with a deeper bed of coffee, you will likely need to increase agitation to achieve an even extraction, which may further the potential for clogging. Grinding slightly coarser will probably be a good idea, with the aim of increasing the water flow and reducing clogging. Be mindful, however, that your optimum brew time will probably be a bit longer than it was before, since matching brew times exactly will likely produce a worse-tasting cup. Consequently, the average extraction yields that you will reach after refining your brew parameters might be slightly lower. If your brews still clog or take too long, you can use less water and then add it back to the brew afterward (i.e., bypass), although this will further reduce your average extraction yield and could produce a less desirable flavor profile. Using a thicker paper filter may be a good strategy when brewing larger doses.
- **Changing the grinder:** If you change your grinder for one with larger burrs or a different geometry that produces either less boulders or less fines, expect your average coffee grinds to look slightly finer by eye. Your optimal brew

time will not necessarily change drastically—although it could. I therefore do not recommend using brew time as any kind of indicator when you change your grinder. The best method I have found is to begin by grinding extremely coarse, such that the grounds visually look too coarse, and your brews are very fast and taste sour and empty. Then, slowly reduce your grind size until you find the brews that taste best. This can be a long process with pourovers. Beginning with a grind that is too fine may produce brews that are harder to diagnose properly. In my experience, the flavor quality did not necessarily decrease linearly as I ground finer when grinding finer than the optimum grind size. Measuring average extraction yields may be useful in determining whether your coffee was underextracted, but I recommend always tasting the coffee first. In more-desperate scenarios, I have sometimes brewed a few coffees at very different grind sizes, blind-tasted them, assigned scores, and then repeated the process until I reached a satisfactory brew. In general, the average extraction yields you obtain at dial-in will tend to be higher if your grinder produces more-even particle size distributions, compared with a grinder that produces less even particle sizes.

- **Sifting boulders:** Sifting coffee boulders will increase the resistance of your coffee bed, and it may require you to grind a bit coarser as a result. If your brews were limited by the amount of fines that your grinder generates, it is possible that sifting out the boulders and grinding coarser will reduce this problem and increase your beverage clarity. The boulders can be reground and put back onto the coffee bed, but this regrinding will generate a larger amount of fines. I do not recommend following any manufacturer's instructions to "simply sift out coffee, do not make up for the change in dose, and do not grind to a different size."
- **Sifting fines:** Sifting fines generally requires a more drastic adaptation in grind size, and it usually makes dialing-in very difficult. This is in part true because it is harder to sift out fines in a repeatable way, but also because it will have a much larger effect on the coffee bed resistance than sifting out boulders would have. We saw in chapters 3 and 4 that the bed resistance is determined by the diameter of the coffee particles at the 10th percentile of the distribution, and removing fines will change this number a lot more than removing boulders would. Expect to need to grind finer and settle on shorter brew times when sifting out fines, because you will drastically reduce the hydraulic resistance of your coffee bed. This does not necessarily mean you will be able to grind so fine that you reach the same brew time as when you do not sift anything out. Doing so may cause you to throw off the taste profile because of the different

ways in which particles of different sizes contribute to flavors (see Chapter 1 for more details). Sifting a large fraction of fines may, however, allow you to obtain cleaner-tasting brews, even when the average extraction yield does not change significantly.

- **Changing the brew ratio:** The most used brew ratios in the specialty coffee community are typically in the range 1:16 to 1:17. Reducing the brew ratio—that is, using less water (e.g., 1:15)—will produce stronger brews with taste profiles that will tend more toward underextraction, whereas increasing it will produce more-extracted and weaker brews. Generally, using a grinder with larger burrs that produces more-even particle size distributions will tend to taste better at larger ratios. For example, I prefer brews made with hand grinders at 1:16 to 1:16.5, those made with the Baratza Forté BG at 1:16.5 to 1:17, and those made with the EG-1 v2 with SSP ultra-low-fines burrs at 1:17. I found that brewing coffees that have a lot of soluble materials (e.g., Kenyans) can taste better at slightly higher ratios (e.g., 1:17.5 or even 1:18, instead of 1:17), whereas those with much less solubles, such as decaffeinated coffees, usually taste better with smaller ratios. I recommend choosing a ratio that seems reasonable for your grinder, dialing-in the grind size without changing the ratio, and then trying different ratios without changing the grind size.
- **Choosing the kettle temperature:** Choosing the right kettle temperature depends on the dripper and brew method you want to use. For most pourover methods, it is important to keep in mind that your slurry temperature will be lower than your kettle temperature, about 5°C (9°F) cooler. For this reason, I tend to prefer drippers such as the Stagg X or plastic V60s, which have a higher slurry temperature, and use a kettle temperature of 99°C (210°F) with all light-roasted coffees. When I taste bitterness in a coffee that is caused by a darker roast, I lower the kettle temperatures by about 3°C (5°F) or until the bitterness is reduced enough that I enjoy the cup. The darkest roasts I have enjoyed have tasted optimal at about 88°C (190°F). I have never found any claims such as "Geshas taste better at lower temperatures" to hold up for me personally, and so far, I have found that cooler kettle temperatures could only really improve the taste of darker roasts.
- **Changing coffee beans:** Using denser, harder beans, such as most Ethiopian coffees, can generate a larger than usual amount of fines, which can lead to longer brew times or even clogging. The first thing I recommend with such coffees is to first try brewing them as usual and then tasting the coffee. I have found that some slow-brewing Ethiopians still tasted amazing at the same grind size, despite the longer brew time. However, often they will taste more

muted or the brew will clog, and in these situations, I recommend adapting strategies to mitigate fines migration. They include the following: not washing out the walls of the paper filter if you use a conical dripper (don't pour directly onto them); grinding coarser; swirling more gently; pouring lower and slower; using thicker filters; and using a smaller coffee dose. In the case of beans that produce much faster flows, it is worth exploring slightly finer grind sizes to determine whether this improves the taste, though in my experience it won't often do so. In such cases, pouring more slowly or using larger doses to increase the brew time may be good approaches. Decaffeinated coffee often requires a coarser grind and a careful management of fines.

- **Assessing the ambient temperature:** If the room temperature where you brew coffee goes up drastically, the amount of fines produced by your grinder may go down, and vice-versa. While I expect this to affect the preparation of espresso more than pourover coffee, you may find that adapting your method or grind size can improve your brews. I recommend adopting similar strategies to those listed above for different coffee bean types.
- **Deciding the number of pours:** Using a larger number of pours will generally increase your brew time, and it has the potential to increase the average extraction yield of your brews because of the longer brew times and the continuous supply of fresh water free of coffee solubles. This can be offset by a reduction in slurry temperature, especially if the level of water becomes lower than about 1 centimeter above the coffee bed, at any point during the brew. I personally use two pours (in addition to the bloom) with plastic V60s and three to five pours with the Stagg X dripper.
- **Changing filters:** When changing filters, pay attention to your drip rate during the brew. Using thinner filters may initially increase the drip rate but cause clogging later, drastically reducing the drip rate. You may need slight adaptations in grind size to make up for this. Very thin and fast-flowing filters may also produce cloudier brews, decreasing taste clarity but increasing the perception of body.
- **Selecting grinder rpm:** Changing your grinder rotation rate (if doing so is possible) will affect the average particle size, so you will likely need to also change the grind size. In my experience, a rotation rate twice as fast requires increasing the burr spacing by about 10%.
- **Changing water:** Keep in mind that preparing coffee in different cities will affect the taste because of different compositions of tap water. Such differences can lead to problems of communication if you brew a coffee using a method suggested by a roaster located in a different city. You can mitigate this by preparing

coffee with custom water compositions, which I suggest you at least try once, in case you are in a city with tap water that produces bad coffee. The details of water composition and how to use custom water are given in Chapter 2.

10.3 An Example V60 Method

Here, I provide an example method of how I would usually prepare a pourover with a plastic V60. I do not think this is the only way to brew a coffee well, but I provide this as a starting point for readers who need it.

- Fold the V60 paper filter across the seam carefully, and press it against a counter or table so that the fold won't lift the filter from the dripper wall.
- Place the dry filter in the dripper and then pour some water into it directly from the faucet so that the filter gets wet entirely and sticks to the walls of the dripper. Let the water drip out; do not pour it out.
- Grind 22 grams of coffee in such a way that it visually resembles coarse sea salt. When the coffee is ready, preheat the dripper with a bit of hot water, let it drain for a moment, and then place the coffee in the dripper. Shake the dripper gently from left to right to flatten the coffee bed, and then make a nest shape in the grounds with your finger (see Chapter 4 for more details). Tare your scale.
- Pour about 70 grams of water for the bloom, beginning at the center of the nest and then going around to evenly wet the coffee bed. Start the timer as soon as water touches the dry coffee. In this step, I usually pour relatively low (just

Folding the seam of a V60 filter. Placing the seam along the corner of a counter is useful to start a straight fold, and then flipping the seam backward and pressing with the fingers ensures that the seam is folded tight and will not loosen during the brew, allowing more water to bypass the coffee bed.

above the dripper) and fast (about 8–10 g/s). Immediately swirl the slurry by moving the dripper around in wide circles. This will capture some coffee fines on the filter walls and flatten the coffee bed.

- At the 45-second mark (assuming water first touched the dry coffee at the zero-second mark), begin the first main pour. The optimum pour rate and height depends on your gooseneck kettle and on the amount of fines your coffee and grinder generate (see Chapter 6). I generally stop this pour at about 200–210 grams of water in total (counting the bloom step). Swirl the slurry gently to flatten the coffee bed immediately after the pour.
- Wait until the level of water drops to just a bit more than 1 centimeter above the coffee bed, and then begin the second main pour. Use a similar pour rate and height, and then swirl the slurry again when the pour is done.
- Note your brew time when the level of water meets the level of the coffee bed.

10.4 An Adapted Method for the Stagg X Dripper

I use a similar method to the one described above for V60 when I brew with the Stagg X. I usually prefer Stagg X brews if they are done properly, but I have found it harder to brew consistently with the Stagg X, in a way that always produces the exact same average extraction yield. I suspect this is due to brew time being more sensitive to filter placement and coffee dose. To simplify things, I have listed below only the differences from the V60 method above:

- I tend to use a slightly larger coffee dose, about 23–25 grams.
- I cut out a tea strainer mesh and placed it at the bottom of the dripper to prevent the filter from collapsing onto the dripper holes and drastically slowing down the drip rate. The mesh must be regularly cleaned with soapy water and thoroughly rinsed to prevent it from slowly clogging with coffee oils, resulting in slower brews yet again. Another option I found works well and does not require regular maintenance is placing a circular food-grade silicon ring at the bottom of the dripper.
- I added a circular base to help swirling; otherwise, swirling evenly is difficult even after removing the rubber base from the Stagg dripper. I used the wooden base of the Hario Olivewood V60, which is pictured at the beginning of this chapter.
- When I brew coffees that yield short brew times (e.g., some South American coffees), after prewetting the filter, I fold the prepleated sides of the filters so that they stick on the dripper walls, to avoid bypass and make the brews slower. Patience is required to do this. Alternatively, prewetting and then forc-

Left: A cutout tea strainer mesh placed at the bottom of the Stagg X dripper can help improve flow, as long as it is cleaned regularly. Right: A food-grade silicon ring can also improve flow, with less maintenance required.

ing a V60 filter into the dripper also works well, but it leads to slightly less repeatable brew times. If you do not fold the prepleated sides so they stick onto the dripper walls, be sure you avoid pouring outside the paper filter.

- I usually divide my main pours into three to five steps with approximately equal water weights, rather than two. I swirl after each one, and I still let them drawdown to a bit more than 1 centimeter of water above the coffee bed.
- I generally prefer slightly longer brew times with the Stagg X dripper (in the range 3:30–5:00) than I do with the V60 (in the range 3:00–4:00).

Left: Folding the prepleated sides of the filter can prevent bypass and make brew times longer. Right: Prewetting and then forcing a V60 filter to fit in the Stagg X can produce great results, especially when combined with a clean tea strainer mesh.

CHAPTER 11

Instruments and Data

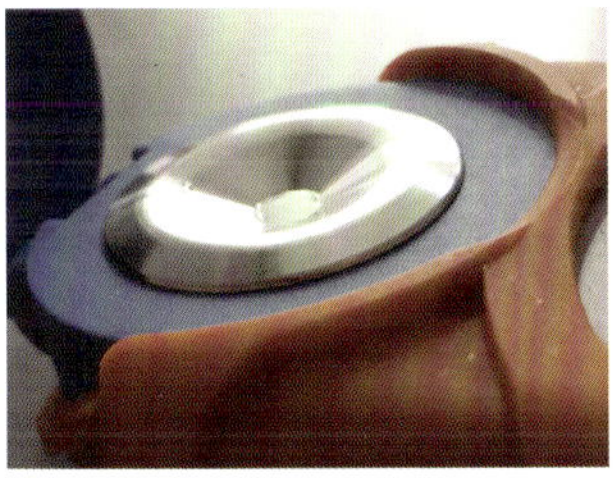

This chapter reviews some of the instruments and data collection methods used in the earlier chapters of the book. I hope readers will forgive me that its content is somewhat more technical; this information will benefit those who would like to either reproduce my results or carry them further.

11.1 The Grind Size Application

Throughout this book and particularly in Chapter 3, I displayed particle size distributions (PSDs) produced by various combinations of grinders, coffees, and situations. Historically in the coffee community, measuring PSDs has been done mostly with efficient but extremely expensive laser diffraction machines. These machines measure the scattering of a point laser source caused by coffee particles in suspension. By utilizing principles of physics, the machines can deduce the diameter of particles passing in front of the laser source.

Another option sometimes used in the coffee community is sifting. A combination of several sieves is typically placed on a shaker. The coffee is shaken in the sieves for a fixed amount of time and subsequently weighed to determine the weight fraction that passed each sieve set. Precision laboratory sieves are still quite expensive and are subject to biases—not all particles will pass through the sieve holes that should be large enough to accommodate them. Results depend on how much shaking is done and on whether the particles adhere to larger particles due for example to static electricity. Some high-end sifting devices use jets of air to counteract these effects, but those devices are prohibitively expensive for home users. Weighing the output precisely can also be a challenge, as some coffee particles can stick to the walls of the sieve sets, and sifting across many dif-

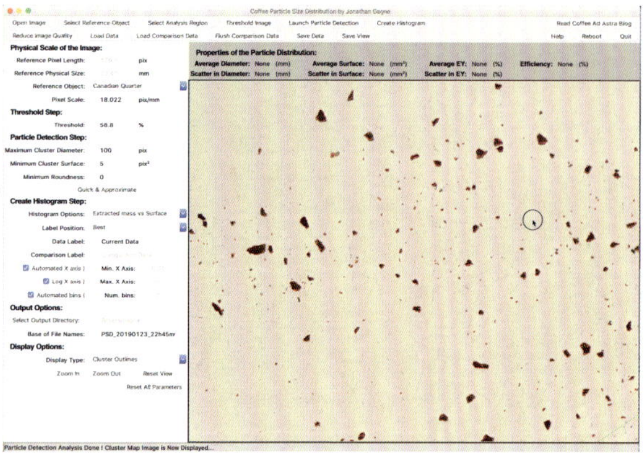

The Grind Size Application

ferent sieve sizes is time consuming and even more prone to the problems mentioned above. Kruve has designed a cheaper sieve set intended for coffee; although the set can be useful for designing custom brew recipes, weighing the weight output accurately for such a large set of sieves is even less practical than it is with laboratory sieves.

About a year ago, when I noticed there was no reliable and affordable way to measure coffee PSDs, I began playing with a microscope imaging software called ImageJ. I used the program to analyze images of ground coffee that I had spread on a flat surface, which allowed me to assess the typical particle sizes. This caused significant interest among coffee professionals, pushing me to write a Python code and create a more user-friendly interface, with additional options useful specifically for analyzing coffee grounds. This led to the construction of the Grind Size Application,[47] which allows users to analyze their own photos of ground coffee. A detailed user manual is available with the application package, and I highly recommend reading it before using the application.

One of its major advantages compared with other methods is that few assumptions need to be made with respect to the particle shapes, because the photos contain the exact shape of each particle, at least across the surface that is observed. Deriving the volume, mass, or total 3D surface of a particle requires some assumptions to be made about the hidden particle depth, but this can be easily done in a more realistic way than assuming spherical shapes. For example, the application assumes that the depth is similar to the smallest diameter seen on the surface, because elongated particles will rarely fall standing up when they are spread on a flat surface. Another advantage of imaging is that particle shapes across the observed surface can also be measured and quantified.

The Grind Size Application comes with its own set of challenges that are distinct from those associated with laser diffraction and sifting. It is much harder, for example, to image coffee fines smaller than 100 μm, because of limited cam-

[47] available at https://coffeeadastra.com/2019/04/07/an-app-to-measure-your-coffee-grind-size-distribution-2/. The source code is also available at https://github.com/jgagneastro/coffeegrindsize.

era resolution and the sticking of fines to larger coffee particles. Another significant challenge is related to the distribution of coffee grounds on a flat surface. For a proper PSD determination, it is crucial that no two particles touch each other; otherwise, the amount of coffee boulders can be vastly overestimated. Some image analysis tools, such as the Watershed algorithm, can help with this problem, but they are far from perfect and will never generate great data from a poorly distributed coffee sample. The best trick I have found to obtain good distribution is to take a pinch of coffee between my index and thumb and spread the coffee by sprinkling it from about 25 centimeters (10 inches) above the flat surface. Controlling how fast you let the coffee fall is crucial to get a good quantity of particles with little clumping.

It may be tempting to spread a very small quantity of particles on a flat surface to avoid clumping, but taken at the extreme this can lead to another problem: it is harder to determine a precise PSD from only a small number of coffee particles. As in all problems of statistics, the larger a sample, the better the output will be, assuming we avoid clumps in the present situation. After I spread the particles on a flat surface, I use a Q-tip® cotton swab dipped in a glass of water to gently remove the coffee chaff. I then use a chopstick to split apart any clumps of coffee particles that occurred despite my precautions. Below, I show an example of how dense I like to spread particles on a white sheet of paper:

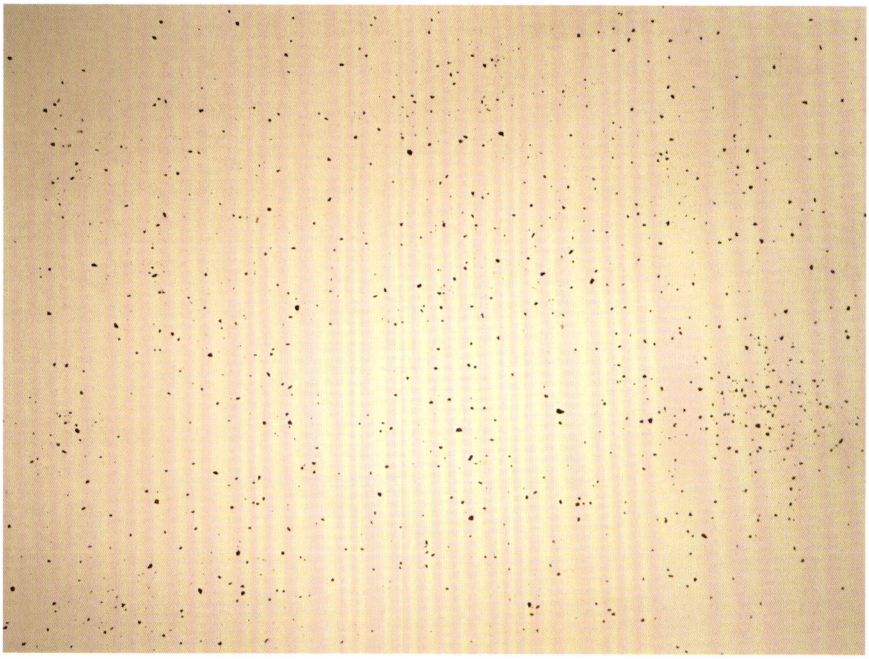

An example photo of coffee grounds obtained with a cell phone and incandescent lamp

I found that resolving subtle PSD differences, such as ~5 μm changes in burr spacing or the effect of coffee bean origin, requires a large number of coffee particles, and as a consequence I often combine several images like the one shown on the previous page to obtain a more accurate PSD. This can be done by exporting each analyzed image as a comma-separated values (CSV) dataset and combining them together, either with an external code or by manually copy-pasting the rows of several files together.[48]

Another important issue to keep in mind when measuring PSDs is that you will only rarely measure a complete sample of ground coffee, as only about 0.2 grams of coffee can fit on a single letter-format white sheet. Because of this, it is crucial that the way you select which particles will be measured does not bias your PSD, either toward encouraging the selection of coffee fines or boulders. As we saw in Chapter 4, any tapping or vibrations will cause your ground coffee sample to sediment across layers of different particle sizes with time. Storing a sample of ground coffee and then sampling only its top or bottom layer is a sure way to bias your PSD measurement significantly. When I take PSD measurements, I place the ground coffee sample in a small, static-free bowl, stir it well with a small spoon in a motion that takes bottom layers and brings them back to the top repeatedly, avoid shaking or tapping the bowl, and pinch a sample of coffee particles by sampling as many vertical layers of coffee particles as possible. Another way to do this is to use the spoon you stirred with to pick up some ground coffee, and sample it by pinching all layers of ground coffee in the spoon itself.

There are yet more effects that can bias the results of a particle size image analysis, including image resolution, lighting, and the determination of pixel scale. Images taken in different lighting environments can change the sensitivity of the software to detect small particles close to the pixel size. As a result, images taken in a very well-lit environment will tend to detect coffee fines more efficiently. A lighting that causes shadows can also be problematic, although in most cases the Grind Size Application is able to distinguish a coffee particle from its shadow. (This can be determined by inspecting the outlines image that the application produces.) Changing camera resolution will also have a similar effect: higher resolutions will allow detecting smaller coffee particles, but keep in mind that high-resolution images will take more time to analyze.

The last problematic effect is the accurate determination of the pixel size in a given image. The Grind Size Application allows a user to do this by placing a coin, ruler, or any other reference object on the image and click on its sides to

[48] Combining CSV files will eventually be implemented as a function within the Grind Size Application.

determine the pixel scale. While this works in most situations and makes the software easier to use, it is not uncommon to see errors of about 0.1–0.2 pixel/mm introduced this way. Measuring small differences in PSDs therefore requires using a stabilized camera that always remains at a fixed distance from the coffee sample and determining the pixel size only once. This way, the PSDs from all images may be slightly inaccurate, but they will be consistent with each other and allow the program to measure small internal differences between the respective PSDs. Keeping a fixed camera distance will also help to maintain similar lighting and resolution, both of which will help with obtaining more accurate results.

An example image of coffee grounds obtained with a scanner at 400 DPI

When he began using the Grind Size Application, Francisco Quijano noticed that using a scanner fixed most of the problems listed above. This is an absolutely brilliant idea, and I found that it makes the results of the application more accurate. The pixel scale can even be calculated from the scanner's DPI, and this voids the need for placing a reference object on each image, making the analysis more straightforward. Users should, however, ensure to turn off any image retouch algorithms (e.g., dust correction) that may be turned on by default with the scanner software. I also recommend using a scanner dedicated for coffee measurements, because some coffee particles can make their way through the cracks and below the scanning glass, which risks damaging the scanner in the long term. I used a relatively cheap scanner (the Epson Perfection V39 with a 400-DPI resolution) for all PSDs that I measured for this book, and I combined three to four images like the one above to generate each PSD.

At the time of the writing of this book, the Grind Size Application displays particle size distributions as a histogram; you must choose a bin size that allows the software to calculate how many particles fall in that bin, along with a measurement error representative of small number statistics. I found that this was a more honest representation of the data than what is typically done with laser diffraction, where a smooth curve is displayed across each bin value, giving a false sense that the continuous PSD was measured, whereas the choice of bin size still matters and affects the measured PSD significantly. Since then, I have begun experimenting with a tool called a kernel density estimate (KDE), which allows a mathematical determination of a continuous PSD with a choice of "bin size" adapted to a sample's properties. The basic idea is to represent the size of each particle with a small normal distribution and sum them collectively to obtain a continuous PSD. The more challenging part is to choose what width each of these normal distributions should have, but a few methods exist to do so (I used Silverman's rule).

Determining a confidence range on the resulting PSD is a bit more complicated when KDEs are involved, and for this reason I used a purely numerical method called a Jackknife Monte Carlo, where a subsample of 80% of the coffee particles is chosen at random to determine a new PSD, and this is repeated a thousand times with different subsets. The resulting thousand PSDs give the reader an idea of how much the results are dependent on the specific sample size, therefore offering an account of measurement errors caused by small number statistics. The KDE features are not yet included in the Grind Size Application, but they will be in a future release.

11.2 Brew Weight Data

Some of the analyses presented in this book relied on the collection of total water weight and beverage weight in real time during the preparation of pour-

Coffee	Roaster	Process	Region	Origin	Varietal	Elevation	Roast Date	Brew Date	Ratio	TIME	BW	TDS	LRR	EY
Heart Ethiopia Abana Sulladjah Batch 47624	Heart	Washed	Gera; Oromia	Ethiopia	Heirlooms	1900-2200	191106	191206	1:17	4:05	321.4	1.47	2.4	21.5
Lüna Coffee Ethiopia Raro Grape Nectar	Lüna	Washed	Uraga; Guji	Ethiopia	Dega	...	190910	191207	1:17	4:14	317.1	1.42	2.6	20.5
Wendelboe Fahem	Tim Wendelboe	Washed	Limmu	Ethiopia	74110; 74148	1900	191022	191208	1:17	4:04	316.6	1.55	2.6	22.4
The Barn Abel Sanchez Mexico	The Barn	Washed	Chiapas	Mexico	Red Bourbon; Garni	1720	191119	191209	1:17	3:11	321.9	1.44	2.4	21.1
Gardelli Alemu Bukato	Gardelli	Natural	Yirgacheffe	Ethiopia	Heirlooms	1900-2100	191101	191211	1:17	4:20	314.5	1.44	2.7	20.6
Heart Kenya Gachuiro AB Batch 48276	Heart	Washed	Nyeri	Kenya	SL28; SL34; Ruiru 1	1915	191210	191214	1:17	3:36	320.6	1.47	2.4	21.4
Heart Kenya Spikes AB Batch 48265	Heart	Washed	Kiambu	Kenya	SL28; SL34	1630	191210	191215	1:17	3:50	320.0	1.47	2.5	21.4
Pilot Tiqiset Wakqo	Pilot	Natural	Yirgacheffe	Ethiopia	Heirlooms	2042	191106	191217	1:17	3:44	320.0	1.40	2.5	20.4
Heart Kenya Spikes AB Batch 48265	Heart	Washed	Kiambu	Kenya	SL28; SL34	1630	191210	191218	1:17	3:37	319.3	1.47	2.5	21.3
Heart Kenya Spikes AB Batch 48265	Heart	Washed	Kiambu	Kenya	SL28; SL34	1630	191210	191220	1:17	3:37	320.9	1.49	2.4	21.7
Heart Kenya Spikes AB Batch 48265	Heart	Washed	Kiambu	Kenya	SL28; SL34	1630	191210	191221	1:17	3:35	320.8	1.47	2.4	21.4
Heart Kenya Spikes AB Batch 48265	Heart	Washed	Kiambu	Kenya	SL28; SL34	1630	191210	191222	1:17	3:51	319.6	1.45	2.5	21.1
Heart Kenya Spikes AB Air1 Batch 48080	Heart	Washed	Kiambu	Kenya	SL28; SL34	1630	191201	191224	1:17	3:32	320.5	1.46	2.4	21.3
Heart Kenya Spikes AB Batch 48265	Heart	Washed	Kiambu	Kenya	SL28; SL34	1630	191210	191229	1:17	4:10	320.8	1.44	2.4	21.0
Heart Ethiopia Worka Sakaro Air 2 Batch 48270	Heart	Washed	Gedeb; Yirgach	Ethiopia	Heirlooms	1800-2000	191210	191229	1:17	3:29	320.6	1.44	2.4	21.0
Passenger Gatomboya AA 2017 Harvest	Passenger	Washed	Nyeri	Kenya	SL28; SL34	1600-1800	190902	191230	1:17	3:40	319.2	1.47	2.5	21.3
Heart Kenya Spikes AB Batch 48265	Heart	Washed	Kiambu	Kenya	SL28; SL34	1630	191210	191231	1:17	3:42	317.3	1.48	2.6	21.3
Howell Gesha Village Estate Oma99 Dec 2017 Harvest	George Howell	Natural	Bench Maji	Ethioopia	Gesha 1931	...	191211	200101	1:17	3:48	320.5	1.42	2.4	20.7
Dragonfly Panama Esmeralda Super Mario 6 Geisha	Dragonfly	Washed	Boquete	Panama	Geisha	1675	190926	200101	1:17	3:48	319.6	1.42	2.5	20.6
Howell Gesha Village Estate Oma99 Dec 2017 Harvest	George Howell	Natural	Bench Maji	Ethioopia	Gesha 1931	...	191211	200103	1:17	4:00	320.8	1.41	2.4	20.6
B&W roasters Kenya Oreti Natural	B&W	Natural	Thika Plateau	Kenya	...	1585	190517	200103	1:17	3:28	322.2	1.49	2.4	21.8
da Matteo Gakuyuini AA Kenya Kaffebox 2019 Dec 4	da Matteo	Washed	Kirinyaga	Kenya	SL28; SL34; Batian	1700	191114	200104	1:17	3:31	322.4	1.43	2.3	21.0
Heart Kenya Gachuiro AB Batch 48276	Heart	Washed	Nyeri	Kenya	SL28; SL34; Ruiru 1	1915	191210	200105	1:17	3:29	318.9	1.44	2.5	20.9
Unido Panama Geisha Finca Deborah Illumination	Unido	Washed Carbonic Mac	...	Panama	Geisha	1920	...	200105	1:17	3:32	323.3	1.37	2.3	20.1
Morgon Testi Adorsi Kaffebox 2019 Dec 5	Morgon	Washed	Aricha; Yirgach	Ethiopia	Heirlooms; Kurume	2150	191118	200106	1:17	3:22	323.9	1.39	2.3	20.5
Pilot Tiqiset Wakqo	Pilot	Natural	Yirgacheffe	Ethiopia	Heirlooms	2042	191106	200111	1:17	3:38	320.6	1.40	2.4	20.4
Heart Ethiopia Abana Sulladjah Batch 47624	Heart	Washed	Gera; Oromia	Ethiopia	Heirlooms	1900-2200	191106	Jan15	1:17	4:03	319.6	1.51	2.5	21.9
Howell Kenya Karatu AA	George Howelll	Washed	Kiambu	Kenya	SL28; SL34	1800	190911	Jan16	1:17	3:33	318.6	1.43	2.5	20.7
Regalia Ethiopia Nano Genji	Regalia	Washed	Jimma; Nano G	Ethiopia	Heirlooms	2200	19042	Jan17	1:17	3:41	316.8	1.51	2.6	21.7
Gardelli El Zafiro Wush Wush	Gardelli	Anaerobic	Popayan	Colombia	Wush Wush	1940-2100	190913	Jan18	1:17	3:18	320.5	1.43	2.4	20.8
Wendelboe 74112 Tatmara Natural	Tim Wendelboe	Natural	Dakiti; Kaffa	Ethiopia	74112	1900-2000	190903	Jan19	1:17	4:15	316.5	1.56	2.6	22.4
Passenger Gesha Village	Passenger	Washed	Gesha	Ethiopia	Gesha	1900-2100	190902	Jan20	1:17	4:19	320.3	1.45	2.4	21.1
Monogram Gesha Village Lot 25	Monogram	Natural	Bench Maji	Ethiopia	Gesha	1969-2069	200101	Jan21	1:17	3:40	318.5	1.53	2.5	22.2
Monogram Gesha Village Lot 25	Monogram	Natural	Bench Maji	Ethiopia	Gesha	1969-2069	200101	Jan21	1:17	3:57	319.9	1.50	2.5	21.8
Monogram Gesha Village Lot 25	Monogram	Natural	Bench Maji	Ethiopia	Gesha	1969-2069	200101	Jan21	1:17	3:52	319.3	1.50	2.5	21.8
Wendelboe 74112 Tatmara Natural	Tim Wendelboe	Natural	Dakiti; Kaffa	Ethiopia	74112	1900-2000	190903	Jan25	1:17	4:03	303.0	1.55	3.4	22.4
Monogram Gesha Village Lot 25	Monogram	Natural	Bench Maji	Ethiopia	Gesha	1969-2069	200101	Jan25	1:17	4:03	304.5	1.50	3.3	21.8
Monogram Gesha Village Lot 25	Monogram	Natural	Bench Maji	Ethiopia	Gesha	1969-2069	200101	Jan26	1:17	3:34	290.0	1.48	4.2	21.5

Some of the brews I logged

The Hario Cube stand and the two Acaia scales connected to Francisco's app on an Android tablet

over coffee. To gather these measurements, I used a Web application[49] developed by Francisco Quijano, connected to an Acaia Pearl scale placed under a Hario Cube Drip Stand and an Acaia Lunar scale placed inside the brew stand and below the server.

The web application receives live readings from both scales; the rate of change of the server scale allows the app to determine the drip rate, and the rate of change of the total water scale allows the app to determine the kettle pour rate. I used a gooseneck tablet stand combined with a cheap Android tablet[50] to connect to both scales and download the data in CSV format. It is important to close the live graphing utility before starting to pour water if the Android tablet has a small amount of system memory or if you plan to brew for more than about 5 min-

A screen capture of the user interface in Francisco's Bluetooth application

---

[49] available at https://pakoquijano.github.io/Acaia_Scale_Web/. This application, developed by Francisco Quijano, is not compatible with Acaia Pearl S scales. Note that official Acaia applications currently cannot connect two scales at once or won't allow the user to export the data as a CSV file. I also found that they crash regularly during use or randomly fail to store data.

[50] Francisco's app works only on Android devices with Google Chrome, or on any laptop with Google Chrome.

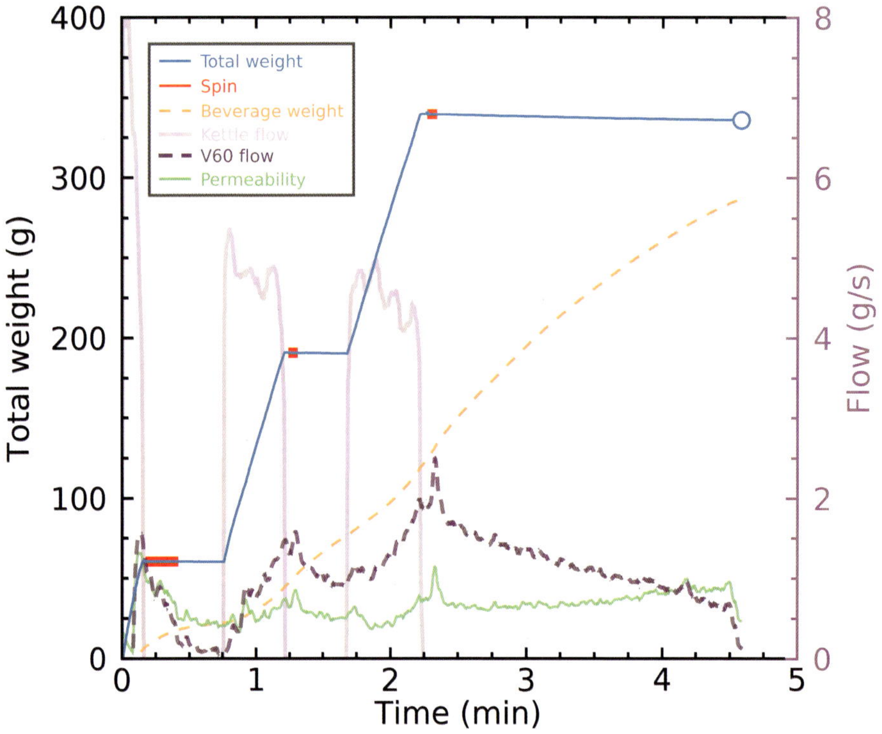

A brew print constructed with custom codes and the CSV data from Francisco's application

utes, because the graphing libraries in the application use a large amount of memory that grows with brew time and risk crashing the app. I used external, custom graphing codes to generate all data related to brew prints that were analyzed in this book.

I place the bead probe just above the spacing in the nest shape before brewing

11.3 Measuring Slurry Temperature

Throughout this book, I measured slurry temperatures by placing a type K beaded thermocouple at the center of the coffee slurry, about 5 millimeters above the dry coffee bed.

The exact positioning of the bead is important because placing it outside the filter or near the dripper

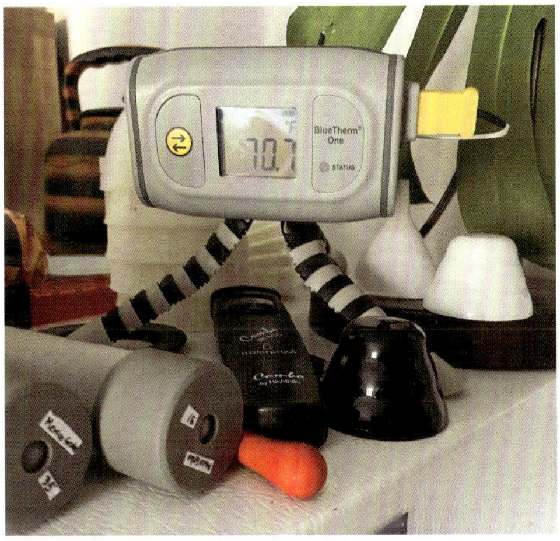

The ThermoWorks BlueTherm® One Bluetooth temperature probe I use to measure my slurry temperature

walls can affect its readings, especially with drippers that are poor thermal insulators. Placing the bead inside the dry coffee bed typically causes channels, and placing it outside the filter near the bottom of a V60 can dramatically affect the amount of water that bypasses the bed.

To perform these temperature measurements, I chose the BlueTherm One temperature probe manufactured by ThermoWorks. It is the only highly calibrated probe (NIST-traceable with a ±0.7°F, or ±0.4°C, accuracy, and a 0.1°F, or 0.06°C, resolution) I could find that comes with a well-designed iOS app allowing me to connect to the probe via Bluetooth and export the data as a CSV file.

11.4 Measuring Roast Color

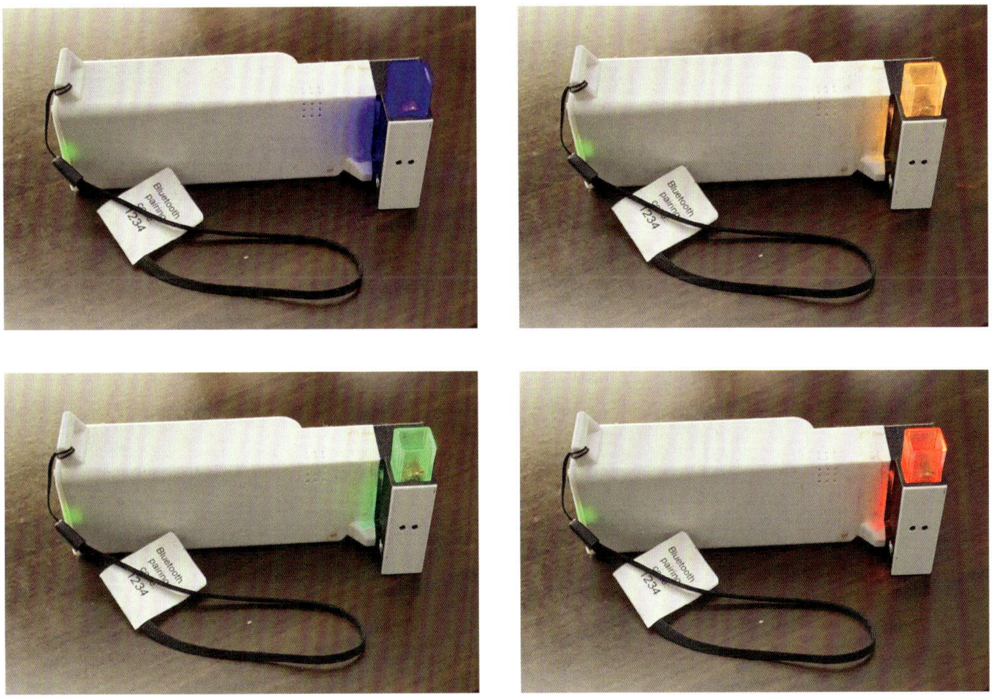

The BioNutrient Meter

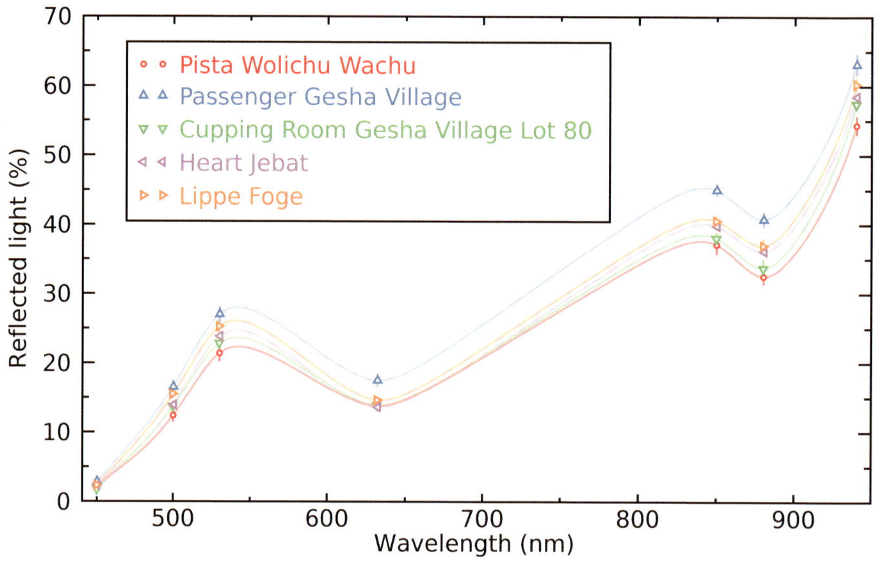

Fraction of light of various wavelengths reflected by ground Ethiopian coffee roasted at different levels

I used a BioNutrient Meter[51] for all roast development measurements presented in Chapter 9 of this book. This device uses two detectors and ten LED lights that range from the UV through visible light and up to the near-infrared (365–940 nm in all). The LEDs are turned on sequentially, and the amount of light reflected on a food sample is measured with a small detector. A black-and-white calibration unit comes with the instrument to correct any drift in LED emission or detector efficiency over time.

I ground each sample of coffee at my usual V60 grind size on the EG-1 v2 grinder with SSP ultra-low-fines burrs and filled the small, square-shaped plastic vial provided with the BioNutrient Meter. I ensured the vial was dry and clean and wiped it up with a microfiber cloth before taking measurements. I wrote a small script for the reflectometer that gathers six measurements to combine them in an average, and I shook the vial (never touching its sides) between each measurement to ensure no effects of voids between the particles affected the readings. I obtained calibration measurements with the black-and-white calibration unit every day I measured coffee colors for calibration.

The figure above shows some example measurements across different Ethiopian coffee beans. Different coffees can produce slightly different slopes, but the main difference is mostly in the average vertical position of the reflectance curve, rather than its shape. Because of this, I chose to represent the bean development level with the average value of the reflectances at 7 wavelengths

[51] https://bionutrient.org/site/bionutrient-meter/

that seemed most stable over time (450, 500, 530, 632, 850, 880, and 940 nm). Because the reflectance values are normalized with respect to the black-and-white calibration unit, black coffee beans would have a reflectance value R near 0%, whereas white beans would have a reflectance value near 100%. I chose to define a roast development level variable D that has a higher value for the darker-roasted beans because this seemed more intuitive, so I defined it as

$$D\,[\%] = 100\% - R\,[\%]$$

such that darker roasts will have a value closer to 100%. The darkest-roasted coffee I have measured, a charcoal-like French-roasted bag of coffee beans I found at a local supermarket, had a roast development value 84.9% (reflectance 15.14 ± 0.07%), and the lightest-roasted coffee I have measured, an Ethiopian Gesha Village roasted by Passenger in September 2019, had a roast development value of 65% (reflectance 35.0 ± 0.6%).

11.5 Analysis of Filter Pores

The microscope imaging analyses of coffee filters that I presented in Chapter 5 were done with a Jiusion 40–1000× magnification USB microscope that uses LED lights to illuminate a sample. I placed the dry paper filters on a flat surface, laid the microscope on top of them, and then adjusted the focus visually. An example of the resulting figure is shown at right.

One of the first steps to perform a pore-size analysis is to determine the pixel scale in this image. To do this, I used a calibration ruler that came with the microscope (below).

I found that the most reliable part of this calibration unit was the grid at the center, because it allowed me to combine multiple measurements of pixel scale across both directions and does not rely on the thickness of lines as they are printed on the calibration unit. A zoom-in on this grid using the USB microscope is shown on the next page.

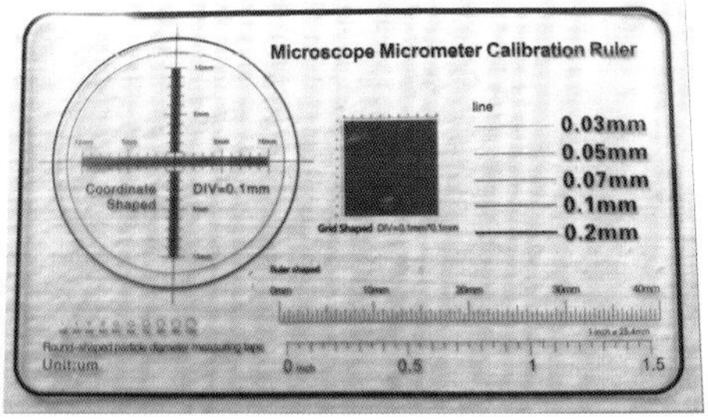

The ruler specifies that the lines are spaced 100 μm apart. While the lines are neither perfectly straight nor uniformly spaced, taking the median value of each column and using an auto-correlation algorithm on the resulting up-and-down pattern allowed me to

determine a pixel scale that averages out these imperfections.

I repeated this analysis both horizontally and vertically on seven distinct microscope images taken at different regions of the grid, and I repositioned the microscope slightly differently on the surface of the calibration ruler to represent manual errors in microscope placement. I found a consistent pixel scale whether I looked in the horizontal (67.59 ± 0.09 μm/pix) or vertical direction (67.54 ± 0.07 μm/pix) and combined them into a single pixel scale of 67.56 ± 0.08 μm/pix, which assumes both directions are independent and that the images are not distorted.

Now that I had the pixel scale in hand, I wrote a code that readjusts the image contrast to obtain a result independent of filter thickness and uses 100% of the red channel added to 50% of the green channel of the color image. This generated a black-and-white image that minimized the impact of slight variations in the brown color of natural, unbleached filters.

The example below shows how a Hario natural filter looks like before (left) and after (right) applying these steps:

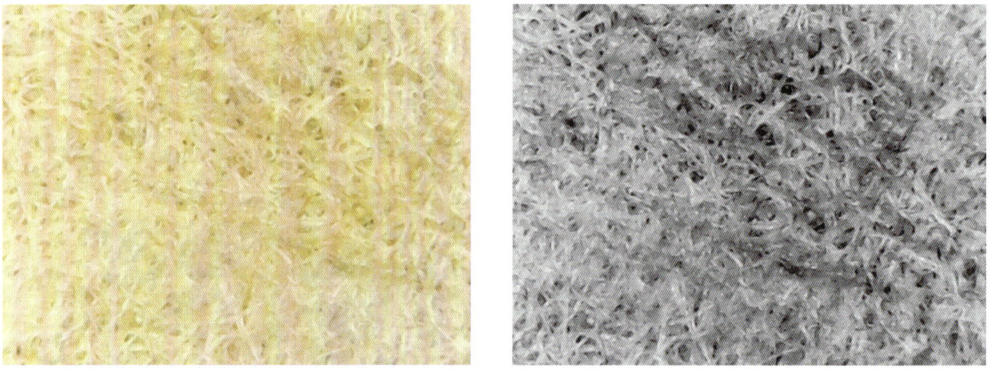

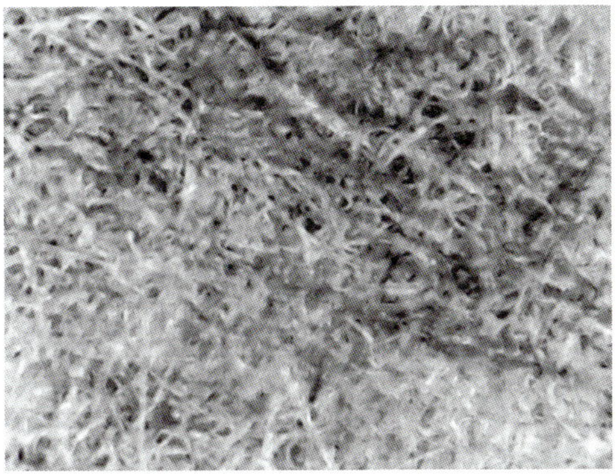

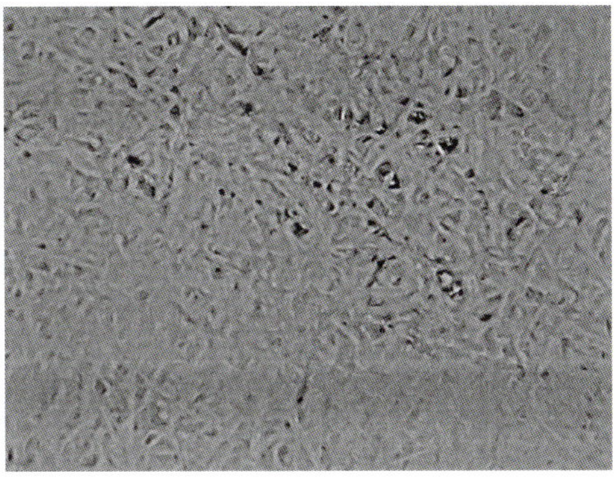

Comparing this last image with the blue channel only (above left) shows how the variations in contrast across the image were indeed reduced by the specific choice of channels shown before.

In order to focus on the actual pores of the coffee filter, I then applied a Butterworth image filter to remove 10% of the largest spatial frequencies of the image. The resulting "pores image" for the same filter is shown above right. I then artificially increased the image resolution by a factor of 20 by interpolation and marked all pixels with flux values below 50% of the full image range as pores. All pixels flagged as pores are shown in red in the resulting image to the left.

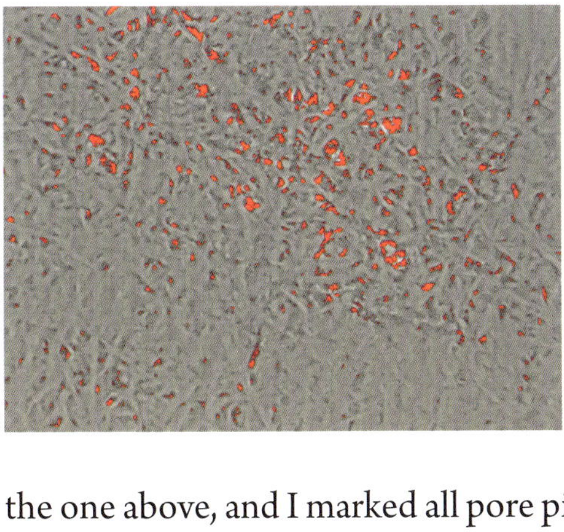

For the next step, I created an image with the same dimensions as the one above, and I marked all pore pixels with a *NaN* ("Not a Number") value and all other pixels that correspond to the filter fibers with a value of zero. I then applied a two-dimensional box-smoothing algorithm that ignores *NaN* values to the image with many box sizes (let's call that box size "x"), with the effect of causing pores marked as *NaN* to creep inward with increasing box size. This is a trick that allowed me to determine at which locations in the smoothed image a square particle of side $2x$ would have been able to pass through a pore.

I then simply counted the fraction $m(x)$ of *NaN* pixels in each smoothed image with a box size x. This function $m(x)$ represents a cumulative distribution of the number of pixels covered by pores of size x or larger divided by the total number of pixels T in the image (with units pixels^2), and it can be mathematically linked to the actual number of pores $N(x)$ per unit pore size, assuming circular pores of area πx^2. This is done by integrating the total area of annuli formed by the outsides of pores with external radii s larger than x, as follows:

$$m(x) = \frac{1}{T} \int_x^\infty \pi (s - x)^2 N(s) \mathrm{d}s$$

Using differential calculus allows us to isolate $N(s)$ by deriving $m(x)$ thrice with respect to the box size x:

$$\frac{\mathrm{d}^3}{\mathrm{d}x^3} m(x) = -\frac{2\pi}{T} N(x)$$

thus, leaving us with

$$N(x) = -\frac{T}{2\pi} \frac{\mathrm{d}^3}{\mathrm{d}x^3} m(x)$$

where $N(x)$ is expressed as a number of pores per unit pore size in pixels. Linking this to a number of pores per unit pore size in microns requires using the pixel scale p (in μm per pixel) that we calculated above:

$$N(r) = -\frac{T}{2\pi p} \frac{\mathrm{d}^3}{\mathrm{d}x^3} m(x) \bigg|_{x=r}$$

This expression can then be used to determine the fractional surface coverage $S(r)$ per unit micron pore size r, with

$$S(r) = \frac{\pi r^2 N(r)}{T p^2}$$

where Tp^2 represents the total image area in physical units. Using our previous expression for $N(r)$ yields

$$S(r) = -\frac{r^2}{2p^3} \frac{\mathrm{d}^3}{\mathrm{d}x^3} m(x) \bigg|_{x=r}$$

Because the microscope has a fixed spatial resolution smaller than our interpolated images, I convolved the resulting $S(r)$ density with a Gaussian filter as wide as the camera resolution[52], assuming it was Nyquist-sampled with a 2.33-pixels wide resolution element. The resulting pore size distributions are displayed in Chapter 5.

---

[52] Convolving an image with a Gaussian filter is a mathematical operation that will effectively blur it.

APPENDIX 1

Mathematical Variables

This section details all the mathematical variables that appear throughout this book, with short descriptions and an example of basic unit dimensions (between parentheses).

Noyes-Whitney Equation

m: Mass of solubles ([mass])

t: Time ([time])

D: Diffusion coefficient ([length]^2/[time])

A: Total surface area of the mass of dry solubles ([length]^2)

C_{sat}: Concentration of water close to the mass of dry solubles ([mass]/[length]^3)

C_{sol}: Concentration of water farther from the mass of dry solubles ([mass]/[length]^3)

L: Distance between the points where C_{sat} and C_{sol} are measured ([length])

Einstein-Smoluchowski Relation

μ_P: Mobility, or the ratio of terminal drift velocity to the corresponding applied force ([length] · [force]$^{-1}$ · [time]$^{-1}$, or [time]/[mass])

k_B: Boltzmann constant, 1.381×10^{-23} joules per kelvin ([entropy], or [energy]/[temperature], or [mass] · [length]^2 · [time]$^{-2}$ · [temperature]$^{-1}$)

T: Temperature ([temperature])

Solution to the Diffusion Equation

c: Concentration ([mass]/[length]^3)

c_0: Maximum concentration ([mass]/[length]^3)

r: Distance from the point of maximum concentration ([length])

D: Diffusion coefficient ([length]^2/[time])

t: Time ([time])

Extraction Yield Equations

LRR: Liquid retained ratio, or the mass of water retained in the spent coffee bed divided by the dry mass of coffee (no physical dimension)

W: Weight of water ([mass])

B: Weight of the final coffee beverage, or brew weight ([mass])

D: Dry mass of coffee, or coffee dose ([mass])

C: Concentration, or total dissolved solids (no physical dimension)

EY: Extraction yield (no physical dimension)

f_{abs}: Weight of water absorbed by coffee particles at the end of a brew divided by the weight of the dry coffee (no physical dimension)

C_{beverage}: Concentration of the coffee beverage, or total dissolved solids, often quoted in % (no physical dimension)

C_{slurry}: Concentration of water retained between coffee particles at the moment when the brew ends, often quoted in % (no physical dimension)

Water Properties

EC: Electric conductivity, often quoted in milliSiemens per centimeter ($[\text{time}]^3 \cdot [\text{current}]^2 \cdot [\text{mass}]^{-1} \cdot [\text{length}]^{-3}$)

TDS: Total dissolved solids, often quoted in % (no physical dimension)

GH: General hardness, often quoted in ppm as $CaCO_3$ (no physical dimension)

KH: Carbonate hardness, or total alkalinity, often quoted in ppm as $CaCO_3$ (no physical dimension)

Particle Size Distributions

k: Hydraulic permeability ($[\text{length}]^2$)

f_{sp}: Function of particle shape and packing acting on the coffee bed permeability (no physical dimension)

D_{10}: Diameter of the 10th percentile coffee particle, by weight ([length])

S_{10}: Surface of the 10th percentile coffee particle, by weight ($[\text{length}]^2$)

V_{10}: Volume of the 10th percentile coffee particle, by weight ($[\text{length}]^3$)

Darcy's Law

Q: Discharge or drip rate ($[\text{length}]^3/[\text{time}]$)

k: Hydraulic permeability ($[\text{length}]^2$)

A: Cross-sectional area ($[\text{length}]^2$)

μ: Dynamic viscosity or absolute viscosity of the fluid, often expressed in units of Pa · s ([mass] · [length]$^{-1}$ · [time]$^{-1}$)

L: Depth of the porous medium ([length])

ρ: Mass density of the fluid ([mass]/[length]^3)

g: Gravitational acceleration ([length]/[time]^2)

h: Total height of the fluid ([length])

Δp: Difference of applied mechanical pressure between the top and bottom of the system ([mass] · [length]$^{-1}$ · [time]$^{-2}$)

K: Hydraulic conductivity ([length]/[time])

R: Hydraulic resistance ([time]/[length])

ν: Kinematic viscosity of the fluid ([length]^2/[time])

Total Brew Time

h_0: Initial height of the water column ([length])

Q_0: Initial drip rate ([length]^3/[time])

t: Time ([time])

T: Total drawdown time ([time])

ρ_c: Mass density of the coffee ([mass]/[length]^3)

ρ_w: Mass density of the water ([mass]/[length]^3)

R: Brew ratio (no physical dimension)

Lucas-Washburn Equation

l: Distance ([length])

κ: Capillary coefficient ([length]/[time]$^{0.5}$)

t: Time ([time])

r: Pore size ([length])

γ: Surface tension of the fluid-to-air interface ([mass]/[time]^2)

μ: Dynamic viscosity or absolute viscosity of the fluid, often expressed in units of Pa · s ([mass] · [length]$^{-1}$ · [time]$^{-1}$)

Fines Migration

F_d: Drag force ([mass] · [length]/[time]^2)

u: Fluid velocity ([length]/[time])

μ: Dynamic viscosity or absolute viscosity of the fluid, often expressed in units of Pa · s ([mass] · [length]$^{-1}$ · [time]$^{-1}$)

r: Radius of the coffee particle ([length])

Q: Discharge or drip rate ([length]^3/[time])

A: Cross-sectional area ([length]^2)

φ: Porosity (no physical dimension)

k: Hydraulic permeability ([length]^2)

L: Depth of the porous medium ([length])

ρ: Mass density of the fluid ([mass]/[length]^3)

g: Gravitational acceleration ([length]/[time]^2)

h: Total height of the fluid ([length])

Δp: Difference of applied mechanical pressure between the top and bottom of the system ([mass] · [length]$^{-1}$ · [time]$^{-2}$)

η: Power of the porosity-to-permeability relation (no physical dimension)

Paper Filters

k_f: Hydraulic permeability of the filter ([length]^2)

A_f: Total exposed surface area of the filter ([length]^2)

L_f: Thickness of the filter ([length])

k_c: Hydraulic permeability of the coffee bed ([length]^2)

A_c: Cross-sectional area of the coffee bed ([length]^2)

L_c: Depth of the coffee bed ([length])

η_f: Number of pores per surface area in the coffee filter ([length]$^{-2}$)

A_p: Cross-sectional surface area of a single filter pore ([length]^2)

Kettle Streams

v_{land}: Velocity of the stream at the landing point ([length]/[time])

v_{kettle}: Velocity of the stream at the point where it exits the kettle spout ([length]/[time])

g: Gravitational acceleration ([length]/[time]^2)

L: Breakup length of the stream ([length])

d_{ag}: Depth of agitation ([length])

s_0: Size of the spray head holes ([length])

Q: Discharge or drip rate ([length]^3/[time])

N: Number of spray head holes (no physical dimension)

APPENDIX 2

Calculations

1. Darcy's Law for Layered Porous Media

When water percolates through more than one porous layer, Darcy's law must be adapted before it can make predictions on flow rate. We saw in Chapter 4 that Darcy's law can be written as

$$Q = \frac{kA}{\mu L}\left(\rho g h - \Delta p\right)$$

We refer the reader to Chapter 4 for a detailed explanation of each term in the equation above, but the schematic shown below can serve as a good visual reminder.

In the present scenario, we want to write a generalized form of Darcy's law that accounts for the two percolation layers, which can have distinct values of hydraulic permeabilities k_1 and k_2, depths L_1 and L_2, and cross-sectional areas A_1 and A_2. The latter will be useful because some drippers may constrain the effective surface of filters if the filter sticks to the dripper walls, as we described in Chapter 7.

The first thing we need to realize to adapt Darcy's law is that water is an incompressible fluid for practical means of coffee brewing. Compressing water requires an impressive amount of force, and even an espresso machine won't come anywhere close to the required pressure drops. Even water at the bottom of the ocean is only compressed by about 2% (Nave 2007). The only way to describe water that is not compressed with a two-layered Darcy's law is to use the same drip rate Q for the top and bottom layers. If the top layer had a faster drip rate, water would be compressed at the interface between the two layers, and vice-versa. Previously, we detailed the two forms of

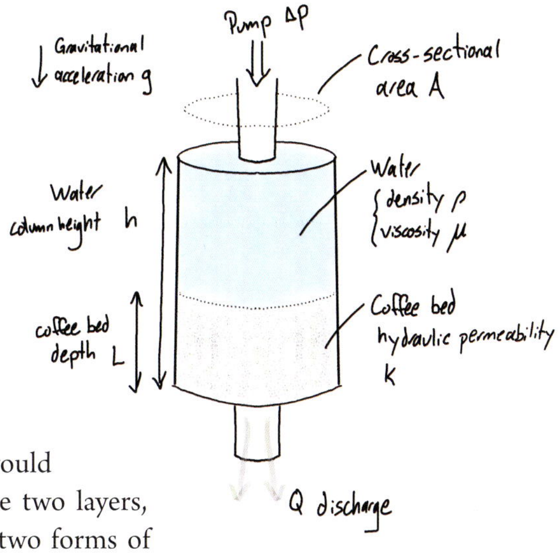

driving forces for the flow as those arising from the weight of water (ρgh) and from any mechanical pump (Δp), but temporarily grouping them together in a new term I call $\Delta p'$ will make this demonstration easier.

These considerations allow us to equate the two distinct Darcy's laws for each porous layer:

$$Q = \frac{k_1 A_1}{\mu L_1} \Delta p_1' = \frac{k_2 A_2}{\mu L_2} \Delta p_2'$$

Now it would be interesting to group these two sides into a single law allowing us to treat both layers as a single system:

$$Q = \left(\frac{kA}{L}\right)_{\text{average}} \frac{\Delta p'}{\mu}$$

In order to do so, we need to use the fact that the pressure drop across all the layers is equal to the sum of individual pressure drops across each layer:

$$\Delta p' = \Delta p_1' + \Delta p_2'$$

The trick is then to solve the first equation for $\Delta p_1'$ and $\Delta p_2'$ individually:

$$\Delta p_1' = \frac{Q \mu L_1}{k_1 A_1}$$

$$\Delta p_2' = \frac{Q \mu L_2}{k_2 A_2}$$

and then insert both of these expressions in the average form of Darcy's law:

$$Q = \left(\frac{kA}{L}\right)_{\text{average}} \left(\frac{Q L_1}{k_1 A_1} + \frac{Q L_2}{k_2 A_2}\right)$$

The discharge Q cancels out, and we are left with

$$\left(\frac{kA}{L}\right)_{\text{average}} = \left(\frac{L_1}{k_1 A_1} + \frac{L_2}{k_2 A_2}\right)^{-1}$$

Technically, this means that the properties of the layered porous media are combined as a *harmonic average*. Using this new relation, we can rewrite the average form of Darcy's law as

$$Q = \left(\frac{L_1}{k_1 A_1} + \frac{L_2}{k_2 A_2}\right)^{-1} \frac{\Delta p'}{\mu}$$

We can also place ourselves back into the situation of a gravity-driven brew, where $\Delta p' = \rho gh$, without an explicit term to account for a mechanical pump:

$$Q = \left(\frac{L_1}{k_1 A_1} + \frac{L_2}{k_2 A_2}\right)^{-1} \frac{\rho gh}{\mu}$$

or alternatively, using the kinematic viscosity ν,

$$Q = \left(\frac{L_1}{k_1 A_1} + \frac{L_2}{k_2 A_2}\right)^{-1} \frac{gh}{\nu}$$

2. Calculation of Total Brew Time

Darcy's law can be used to predict the total brew time in a cylindrical dripper with a single, fast pour. In the absence of a mechanical pump, gravity is the lone driving force behind the flow of water, and Darcy's law becomes

$$Q = \frac{kAgh}{\nu L}$$

We are looking for a solution of how the water column height h changes over time because of its dripping out, and to use this effect to determine how long it will take for the height of the water column to reach the height of the coffee bed ($h = L$), which is how we defined the end of a brew in this book.

The first thing we need to realize to make a connection between h and Q is that Q represents a decrease of water volume over time:

$$Q(t) = -\frac{\mathrm{d}V}{\mathrm{d}t} = -\frac{\mathrm{d}}{\mathrm{d}t}\Big(A \cdot h(t)\Big)$$

we can replace $Q(t)$ in the expression above using Darcy's law and obtain a linear differential equation that describes the evolution of $h(t)$:

$$\frac{\mathrm{d}h}{\mathrm{d}t} = -\frac{kg}{\nu L}h(t)$$

The solution to this differential equation has the form of an exponentially decreasing column height:

$$h(t) = H_0 \exp\Big(-\frac{kg}{\nu L}t\Big)$$

where H_0 is the initial water column height. Posing the constraint $h(T) = L$, where T is the total brew time, and solving for T leads to

$$T = \frac{\nu L}{kg}\ln\left(\frac{H_0}{L}\right)$$

We can also express the total brew time as a function of the brew ratio R, which can be more directly interpreted than the initial water column height H_0. The brew ratio R can be written as

$$R = \frac{\rho A H_0}{\rho_c V_c}$$

where ρ is the mass density of water, ρ_c is the mass density of ground coffee, V_c is the volume of the coffee dose, and the product AH_0 is the volume of the initial water column. Noting that we have $V_c = AL$ by geometry, we can rewrite H_0 as

$$H_0 = \frac{\rho_c L}{\rho}R$$

We can then rewrite our final equation for the total brew time:

$$T = \frac{\nu L}{kg}\ln\left(\frac{\rho_c R}{\rho}\right)$$

It is possible to do a similar calculation for a conical dripper using the modified form of Darcy's law derived further below, but the solution to the differential equation for $h(t)$ becomes transcending and therefore necessitates a numerical solution.

3. Darcy's Law for a Conical Dripper Without Bypass

In order to modify Darcy's law for different dripper geometries, we will need to use differential calculus; only readers comfortable will this branch of mathematics will therefore benefit from this explanation. The first step in doing this is to write a form of Darcy's law that describes an infinitesimally thin tube of fluid and porous medium, with a cross-sectional area dA and a small contribution dQ to the total drip rate:

$$\mathrm{d}Q = \frac{kgh}{L\nu}\mathrm{d}A$$

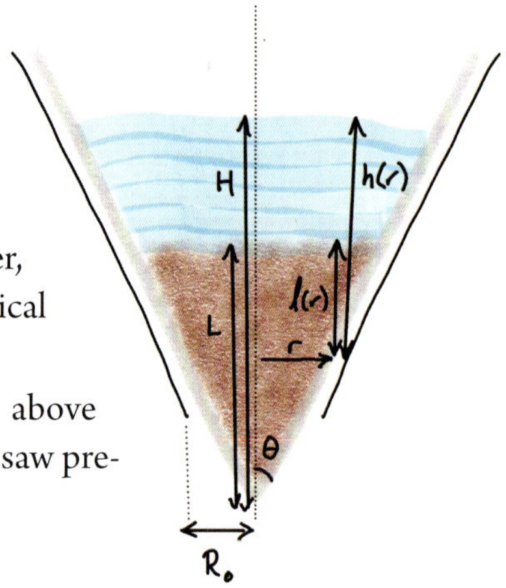

If we assume no water bypasses the coffee bed, we can simply calculate the total flow rate by taking the sum of all infinitesimally thin vertical tubes of water, coffee, and filter, over the opening hole at the bottom of the dripper, as demonstrated in the schematic at right. The mathematical operation that allows us to do this is an integral.

Adapting the infinitesimal version of Darcy's law above for such a vertical tube using the demonstration that we saw previously for layered percolation media yields:

$$\mathrm{d}Q = \frac{gh(r)}{\nu}\left(\frac{l(r)}{k} + \frac{1}{\beta}\right)^{-1}\mathrm{d}A$$

where $l(r)$ is the thickness of the coffee bed at radial distance r from the center of the coffee dose, k is the hydraulic permeability of the coffee bed, and β is the ratio of hydraulic permeability over the thickness of the coffee filter.

If we define θ as the half-angle of the dripper walls (30° for the V60 dripper) and L as the total depth of the coffee bed, we can use basic trigonometry to determine the radial profile of coffee bed thickness:

$$l(r) = L - \frac{r}{\tan\theta}$$

A similar geometric consideration allows us to write the radial profile of the water column height:

$$h(r) = H - \frac{r}{\tan\theta}$$

where H is the total water column height measured at the center of the dripper. Considering the infinitesimal element of area dA in polar coordinates (r, φ):

$$\mathrm{d}A = r\mathrm{d}r\mathrm{d}\varphi$$

we can rewrite and simplify the infinitesimal Darcy law given above to:

$$\mathrm{d}Q = \frac{g}{\nu}\left(H - \frac{r}{\tan\theta}\right)\left(\frac{k}{L - r/\tan\theta + k/\beta}\right) r\mathrm{d}r\mathrm{d}\varphi$$

Taking the integral over the range $[0, 2\pi[$ in φ and over $[0, R_0]$ in r, where R_0 is the radius of the dripper hole, yields:

$$Q = \frac{2\pi g k}{\nu}\left(H\int_0^{R_0}\frac{r}{L + k/\beta - r/\tan\theta}\mathrm{d}r - \frac{1}{\tan\theta}\int_0^{R_0}\frac{r^2}{L + k/\beta - r/\tan\theta}\mathrm{d}r\right)$$

A change of variables can be used to express the integrals in a simpler form:

$$Q = \frac{2\pi g k \tan^2\theta}{\nu}\left(H\int_0^{H_0}\frac{z}{\gamma - z}\mathrm{d}z - \int_0^{H_0}\frac{z^2}{\gamma - z}\mathrm{d}z\right)$$

where $z = r/\tan\theta$, $H_0 = R_0/\tan\theta$ and $\gamma = L + k/\beta$. Solving both integrals yields:

$$\int_0^{H_0}\frac{z}{\gamma - z}\mathrm{d}z = -H_0 - \gamma\ln\left(1 - \frac{H_0}{\gamma}\right)$$

$$\int_0^{H_0}\frac{z^2}{\gamma - z}\mathrm{d}z = -\gamma H_0 - \frac{H_0^2}{2} - \gamma^2\ln\left(1 - \frac{H_0}{\gamma}\right)$$

These solutions allow us to re-write and simplify our expression for the discharge Q:

$$Q = \frac{2\pi g k \tan^2\theta}{\nu}\left(\frac{H_0^2}{2} + H_0(\gamma - H) + \gamma(\gamma - H)\ln\left(1 - \frac{H_0}{\gamma}\right)\right)$$

$$\gamma = L + \frac{k}{\beta}$$

4. Darcy's Law for a Conical Dripper with Bypass

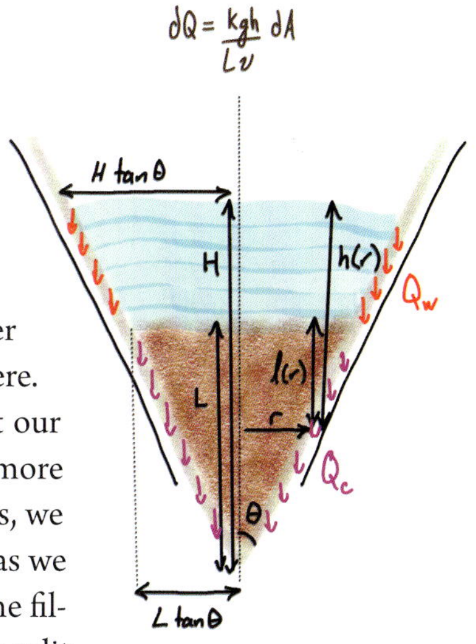

In this section, we derive a modified Darcy's law for a conical dripper that allows for a maximal amount of bypass, meaning that we assume there are either no dripper walls, or the filter is well separated from the walls everywhere.

To account for the effect of bypass, we must adapt our previous calculations for the fact that water can now take more than one type of paths through the coffee bed. To do this, we return to the same infinitesimal equation of Darcy's law as we did before, and we assume that water can pass through the filter even beyond the dripper hole ($r > R_0$), requiring us to split the total discharge Q across two regimes.

The first regime corresponds to regions where water has to cross both a path of the coffee bed and the filter ($r < L\tan\theta$), contributing to a discharge Q_c, and the second regime corresponds to regions where water only has to cross the paper filter ($r \geq L\tan\theta$), contributing to a discharge Q_w, the sum of which corresponds to the total discharge:

$$Q = Q_c + Q_w$$

The first regime has exactly the same integrals and solution that we derived before, for the case without bypass, except that we need to extend the upper bounds of the integral from $r = R_0 = H_0 \tan\theta$, up to $r = L \tan\theta$. The effect will be that every instance of H_0 in the previous solution must be replaced by L. We therefore have:

$$Q_\mathrm{c} = \frac{2\pi g k \tan^2\theta}{\nu}\left(\frac{L^2}{2} + L(\gamma - H) + \gamma(\gamma - H)\ln\left(1 - \frac{L}{\gamma}\right)\right)$$

The second regime where water passes only through the filter has a simpler infinitesimal version of Darcy's law:

$$\mathrm{d}Q_\mathrm{w} = \frac{g}{\nu} h(r)\beta \mathrm{d}A$$

Using our previous definitions of $h(r)$ and $\mathrm{d}A$, solving the integral over $\mathrm{d}\varphi$ and changing the integration variable from r to z yields:

$$Q_\mathrm{w} = \frac{2\pi g\beta \tan^2\theta}{\nu}\int_L^H z\,(H - z)\,\mathrm{d}z$$

Solving this integral and simplifying allows us to write our final form for Q_w:

$$Q_\mathrm{w} = \frac{2\pi g\beta \tan^2\theta}{\nu}\left(\frac{H^3}{6} - \frac{HL^2}{2} + \frac{L^3}{3}\right)$$

Taking the sum of Q_c and Q_w and further simplifying leads to:

$$Q = \frac{2\pi g \tan^2\theta}{\nu}\left[k\left(\frac{L^2}{2} + L(\gamma - H) + \gamma(\gamma - H)\ln\left(1 - \frac{L}{\gamma}\right)\right) + \beta\left(\frac{H^3}{6} - \frac{HL^2}{2} + \frac{L^3}{3}\right)\right]$$

One of the major differences between this form of Darcy's law and that of the conical or cylindrical forms without bypass is the fact that the drip rate Q now depends on the third power of the water column height H, rather than being linear with H. Intuitively, this is caused by the increased surface of contact between the paper filter and water when the water column is taller, which allows for more water to bypass the coffee bed completely.

Realistic cases of a V60 with ridges will be in between the two cases we studied here, because water will only bypass between the ridges and the paper filter may temporarily adhere to the filter walls even between the ridges. However, the random nature of this process of filter adhesion prevents the derivation of a general equation for this intermediate case.

In principle, it would be possible to adapt the bypass case with an adhesion factor that reduces the value of β only in the range $R_0 < r < L \tan\theta$ and to determine this adhesion factor numerically at each moment in time during a brew, but doing so would require assuming that the coffee bed permeability k does not also simultaneously change during the brew, which is not realistic because of fines migration and channeling. Additional data would therefore be needed to separately determine the exact filter adhesion and coffee bed permeability at every moment in time during a brew.

5. General Equation for the Average Extraction Yield

By the definition we adopted in Chapter 1, the average extraction yield most useful to estimate the flavor profile in a cup of coffee is the total mass of solubles that left the ground coffee particles divided by the total dry mass of the coffee dose (D), at the moment where the brew ended. As we saw before, the reasoning behind this argument is that the profile of chemicals extracting at every moment in time changes as the total mass of extracted chemicals increases. This means that considering the solubles left in interstitial water helps us to assess the past extraction history of the brew, and thus the profile of chemicals in the cup.

We can divide that mass of extracted solubles in two parts: those that made their way into the cup of coffee (M_{bev}) and those that remained in the interstitial water retained in the spent coffee bed (M_{ret}). We thus have the following equation for the average extraction yield E:

$$E = \frac{M_{\text{bev}} + M_{\text{ret}}}{D}$$

Assuming that we measured the total dissolved solids concentration of the beverage (C_{bev}) and estimated the concentration of the interstitial water by measuring the concentration of the last few drops (C_{last}) using a refractometer, we can deduce the mass of coffee solubles in the beverage:

$$M_{\text{bev}} = B \cdot C_{\text{bev}}$$

where B is the total mass of the coffee beverage. Both C_{bev} and C_{last} are defined here as taking values between zero and 1 (e.g., 1.4% TDS would correspond to a value 1.4/100 = 0.014). Deducing the mass of coffee solubles retained in the interstitial water is slightly less straightforward: we must first consider the mass of the interstitial water at the end of the brew (W_{int}) and use the definition of a concentration as a mass ratio:

$$C_{\text{last}} = \frac{M_{\text{ret}}}{M_{\text{ret}} + W_{\text{int}}}$$

As mentioned before, we have assumed in this equation that the concentration of the last few drops is a good approximation of the concentration of the interstitial water retained in the spent coffee bed at the moment where the brew ended. We can reverse the equation above by solving for M_{ret}:

$$M_{\text{ret}} = W_{\text{int}} \cdot \frac{C_{\text{last}}}{1 - C_{\text{last}}}$$

One complication with the equation above is that we don't have an easy way to measure W_{int} directly, especially considering that some more water is also retained in the spent coffee bed as water absorbed inside coffee particles (W_{abs}). We can link the total retained water with the mass of the coffee beverage B, the mass of extracted coffee solubles in the cup (M_{bev}) and the total mass of water that was poured over the coffee W:

$$B = W - W_{\text{int}} - W_{\text{abs}} + M_{\text{bev}}$$

Solving for W_{int} and inserting this expression in the equation for M_{ret} yields:

$$M_{\text{ret}} = \frac{C_{\text{last}}}{1 - C_{\text{last}}} \left(W - B + M_{\text{bev}} - W_{\text{abs}}\right)$$

We can also insert here our equation for M_{bev} from before and obtain:

$$M_{\text{ret}} = \frac{C_{\text{last}}}{1 - C_{\text{last}}} \Big(W - B\left(1 - C_{\text{bev}}\right) - W_{\text{abs}}\Big)$$

Now that we have linked both M_{ret} and M_{bev} to other relevant parameters, we can go back and combine them in our equation for the average extraction yield E, and simplify it with some straightforward algebra:

$$E = \frac{B \cdot C_{\text{bev}}}{D} + \frac{C_{\text{last}}}{1 - C_{\text{last}}} \frac{W - W_{\text{abs}}}{D} - \frac{C_{\text{last}}\left(1 - C_{\text{bev}}\right)}{1 - C_{\text{last}}} \frac{B}{D}$$

$$E = \frac{C_{\text{bev}} - C_{\text{last}}}{1 - C_{\text{last}}} \frac{B}{D} + \frac{C_{\text{last}}}{1 - C_{\text{last}}} \frac{W - W_{\text{abs}}}{D}$$

Now we can introduce f_{abs}, a quantity analogous to the liquid retained ratio, but that counts only the fraction of water that was absorbed by the coffee particles:

$$f_{\text{abs}} = \frac{W_{\text{abs}}}{D}$$

and use it to obtain the final form of our equation for the average extraction yield:

$$E = \frac{C_{\text{bev}} - C_{\text{last}}}{1 - C_{\text{last}}} \frac{B}{D} + \frac{C_{\text{last}}}{1 - C_{\text{last}}} \left(\frac{W}{D} - f_{\text{abs}}\right)$$

While f_{abs} cannot be measured directly and may depend on grinder and coffee beans, it seems to be independent of grind size in my experience, and can be estimated with an AeroPress, which can remove most of the interstitial water if the plunger is pressed far enough.

References

Aharonov, E., E. Tenthorey, and C. H. Scholz. 1998. "Precipitation Sealing and Diagenesis: 2. Theoretical Analysis." *Journal of Geophysical Research Solid Earth* 103, no. B10 (October): 23969–23981. https://doi.org/10.1029/98JB02230.

Ahmadi, S., and F. Sefidvash. 2018. "Study of Pressure Drop in Fixed Bed Reactor Using a Computational Fluid Dynamics (CFD) Code." *ChemEngineering* 2, no. 2 (April): 14. https://doi.org/10.3390/chemengineering2020014.

Aksu, I., E. Bazilevskaya, and Z. Karpyn. 2015. "Swelling of Clay Minerals in Unconsolidated Porous Media and Its Impact on Permeability." *GeoResJ* 7 (September): 1–13. https://doi.org/10.1016/j.grj.2015.02.003.

Anderson, K. "Static in Your Grinding—It's Complex!" *Baratza* (blog). March 7, 2014. https://baratza.com/static-in-your-grinding-its-complex/.

Anderson, W., C. Cooper, J. Fredland, M. Lemieux, M. Pabst, D. Pezza, and H. Van Aller. 2007. *Technical Manual: Plastic Pipe Used in Embankment Dams.* Washington, D.C.: FEMA.

Anese, M., L. Manzocco, and N. Nicolli. 2006. "Modeling the Secondary Shelf Life of Ground Roasted Coffee." *Journal of Agricultural and Food Chemistry* 54, no. 15 (June): 5571–5576. https://doi.org/10.1021/jf060204k.

Baird, R. B., A. D. Eaton, and E. W. Rice, eds. 2012. *Standard Methods for the Examination of Water and Wastewater.* Washington, D. C.: American Public Health Association.

Baggenstoss, J., D. Thomann, R. Perren, and F. Escher. 2010. "Aroma Recovery from Roasted Coffee by Wet Grinding." *Journal of Food Science* 75, no. 9 (October): 697–702. https://doi.org/10.1111/j.1750-3841.2010.01822.x.

Baylar, A., and M. E. Emiroglu. 2004. "An Experimental Study of Air Entrainment and Oxygen Transfer at a Water Jet from a Nozzle with Air Holes." *Water Environment Research* 76, no. 3 (May): 231–7. https://doi.org/10.2175/106143004X141780.

Bear, J. 1988. *Dynamics of Fluids in Porous Media.* New York: Dover.

Bekele, G., and T. Hill. 2018. *A Reference Guide to Ethiopian Coffee Varieties.* Durham: Counter Culture Coffee.

Bergendahl, J. A., and D. Grasso. 2003. "Mechanistic Basis for Particle Detachment from Granular Media." *Environmental Science & Technology* 37, no. 10 (April): 2317–22. https://doi.org/10.1021/es0209316.

Bernabé, Y., and B. Evans. 2003. "Permeability–Porosity Relationships in Rocks Subjected to Various Evolution Processes." *Pure and Applied Geophysics* 160, no. 5 (May): 937–60. https://doi.org/10.1007/PL00012574.

Bertola, V., and G. Brenn. 2019. "Drops and Sprays," in *Transport Phenomena Across Interfaces of Complex Fluids*, ed. T. Burghelea and V. Bertola, 293–360. Cham: Springer.

Bollaert, E. *Transient Water Pressures in Joints and Formation of Rock Scour Due to High-Velocity Jet Impact.* PhD diss., École polytechnique fédérale de Lausanne, 2002.

Bollaert, E., and A. J. Schleiss. 2003. "Scour of Rock Due to the Impact of Plunging High Velocity Jets. Part II: Experimental Results of Dynamic Pressures at Pool Bottoms and in One- and Two-Dimensional Closed End Rock Joints." *Journal of Hydraulic Research* 41, no. 5: 465–80. https://doi.org/10.1080/00221680309499992.

Brown, R. 2018. *Dear Coffee Buyer.* Self-published.

Burke, R. "Coffee Beans." (blog). November 27, 2018. http://www2.optics.rochester.edu/workgroups/cml/opt307/spr16/beckah/index.htm.

Butterworth, M. "Grinding for a Purpose." *Freshcup* (blog). March 7, 2016. https://www.freshcup.com/grinding-for-a-purpose/.

Callister W. D. J., and D. G. Rethwisch. 2018. *Materials Science and Engineering: An Introduction,* 10th ed. New Jersey: Wiley.

Cameron, M. "The Resistance of Heat." (blog). September 25, 2016. https://strivefortone.com/2016/09/25/the-resistance-of-heat/.

Cameron, M. I., D. Morisco, D. Hofstetter, E. Uman, J. Wilkinson, Z. Kennedy, S. Fontenot, W. Lee, C. Hendon, and Foster. J. 2020. "Systematically Improving Espresso: Insights from Mathematical Modeling and Experiment." *Matter* 2, no. 3 (January): 631–48. https://doi.org/10.1016/j.matt.2019.12.019.

Cappuccio, R., and F. S. Liverani. 1999. *18th International Scientific Colloquium on Coffee: Computer Simulation as a Tool to Model Coffee Brewing Cellular Automata for Percolation Processes; 2D and 3D Techniques for Fluid-Dynamic Simulations, Helsinki, Finland.* Paris: ASIC—Association for Science and Information on Coffee Proceedings.

Colón, C. F. J., E. H. Oelkers, and J. Schott. 2004. "Experimental Investigation of the Effect of Dissolution on Sandstone Permeability, Porosity, and Reactive Surface Area." *Geochimica et Cosmochimica Acta* 68, no. 4 (February): 805–17. https://doi.org/10.1016/j.gca.2003.06.002.

Colonna-Dashwood, M. "Maxwell Colonna-Dashwood—Water Chemistry and Its Impact on Coffee Flavour." YouTube video, January 5, 2016. https://www.youtube.com/watch?v=VAwxrxPFEMg.

Colonna-Dashwood, M., and C. Hendon. 2015. *Water for Coffee.* Self-published.

Corrochano, B. R., J. R. Melrose, A. C. Bentley, P. J. Fryer, and S. Bakalis. 2015. "A New Methodology to Estimate the Steady-State Permeability of Roast and Ground Coffee in Packed Beds." *Journal of Food Engineering* 150 (April): 106–16. https://doi.org/10.1016/j.jfoodeng.2014.11.006.

Crowe, C. T., ed. 2005. *Multiphase Flow Handbook.* Boca Raton: CRC Press.

Dapp, W. B., and M. H. Müser. 2016. "Fluid Leakage Near the Percolation Threshold." *Scientific Reports* 6 (February): 19513. https://doi.org/10.1038/srep19513.
de Maria, C. A. B., L. C. Trugo, R. F. A. Moreira, and C. C. Werneck. 1994. "Composition of Green Coffee Fractions and their Contribution to the Volatile Profile Formed During Roasting." *Food Chemistry* 50 no. 2: 141–45. https://doi.org/10.1016/0308-8146(94)90111-2.
Duarte, R., A. Pinheiro, and A. J. Schleiss. 2016. "An Enhanced Physically Based Scour Model for Considering Jet Air Entrainment." *Engineering* 2, no. 3 (September): 294–301. https://doi.org/10.1016/J.ENG.2016.03.003.
Duez, C., C. Ybert, C. Clanet, and L. Bocquet. 2010. "Beating the Teapot Effect." *Physical Review Letters* 104, no. 8 (February): 84503. https://doi.org/10.1103/PhysRevLett.104.084503.
Easthope, A. "Brew Temperature and Its Effects on Espresso." *Five Senses* (blog). April 8, 2015. https://www.fivesenses.com.au/blog/brew-temperature-and-its-effects-on-espresso/.
Ellero, M., and L. Navarini. 2019. "Mesoscopic Modelling and Simulation of Espresso Coffee Extraction." *Journal of Food Engineering* 263 (December): 181–94. https://doi.org/10.1016/j.jfoodeng.2019.05.038.
Farah, A. 2019. *Coffee: Production, Quality and Chemistry.* Croydon: Royal Society of Chemistry.
Folmer, B., ed. 2017. *The Craft and Science of Coffee.* Amsterdam: Academic Press.
Georgiadis, A., G. Maitland, J. P. M. Trusler, and A. Bismarck. 2010. "Interfacial Tension Measurements of the (H_2O + CO_2) System at Elevated Pressures and Temperatures." *Journal of Chemical & Engineering Data* 55, no. 10 (May): 4168–75. https://doi.org/10.1021/je100198g.
Ghanbarian, B., A. G. Hunt, and H. Daigle. 2016. "Fluid Flow in Porous Media with Rough Pore-Solid Interface." *Water Resources Research* 52, no. 3 (February): 2045–58. https://doi.org/10.1002/2015WR017857.
Grathwohl, P. 1998. *Diffusion in Natural Porous Media.* New York: Springer US.
Guilizzoni, M., M. Santini, and S. Fest-Santini. 2019. "Synchronized Multiple Drop Impacts into a Deep Pool." *Fluids* 4, no. 3 (July): 141. https://doi.org/10.3390/fluids4030141.
Hanžič, L., L. Kosec, and I. Anželc. 2010. "Capillary Absorption in Concrete and the Lucas–Washburn Equation." *Cement and Concrete Composites* 32, no. 1 (January): 84–91. https://doi.org/10.1016/j.cemconcomp.2009.10.005.
Haraldsson, H. O. *Breakup of Jet and Drops During Premixing Phase of Fuel Coolant Interactions.* PhD diss., KTH Royal Institute of Technology, 2000.
Hazen, A. 1911. "Discussion of Dams on Sand Foundations." *Transaction of the American Society of Civil Engineers* 73, no. 3: 199–203.
Hoffmann, A. C., and H. J. Finkers. 1995. "A Relation for the Void Fraction of Randomly Packed Particle Beds." *Powder Technology* 82, no. 2 (February): 197–203. https://doi.org/10.1016/0032-5910(94)02910-G.

Hoffmann, J. "Crema is Rubbish." *jimseven* (blog). July 6, 2009. https://jimseven.com/2009/07/06/video-1-crema/.

Hoffmann, J. 2014. *The World Atlas of Coffee: From Beans to Brewing—Coffees Explored, Explained and Enjoyed.* London: Octopus Book Publishing Group Ltd.

Hoffmann, J. "Weird Coffee Science—The Hard Bloom." YouTube video, July 22, 2019. https://www.youtube.com/watch?v=MpKEdZjpqXM.

Hoffmann, J. "Weird Coffee Science—Microwave Your Coffee Beans." YouTube video, August 15, 2019. https://www.youtube.com/watch?v=yXCboVCDHLw.

Hoffmann, J. "Weird Coffee Science—High Humidity Coffee Storage." YouTube video, January 20, 2020. https://www.youtube.com/watch?v=J7xEbf1mkm0.

Hoffmann, J. "Weird Coffee Science—Humid Coffee Gets Weirder (Part 2)." YouTube video, April 28, 2020. https://www.youtube.com/watch?v=BcXK1MZJDAM.

Holzbecher, E., and H. Oehlmann. 2012. *Proceedings of the 2012 COMSOL Conference: Comparison of Heat and Mass Transport at the Micro-Scale, Milan, Italy, October 10-12, 2012.* Stockholm: COMSOL.

Hommel, J., E. Coltman, and H. Class. 2018. "Porosity–Permeability Relations for Evolving Pore Space: A Review with a Focus on (Bio-)geochemically Altered Porous Media." *Transport in Porous Media* 124, no. 2 (May): 589–629. https://doi.org/10.1007/s11242-018-1086-2.

Hunt, A. G., and M. Manga. 2003. "Effects of Bubbles on the Hydraulic Conductivity of Porous Materials—Theoretical Results." *Transport in Porous Media* 52, no. 1 (July): 51–65. https://doi.org/10.1023/A:1022312927606.

Illy, A., and R. Viani, eds. 2004. *Espresso Coffee: The Science of Quality.* Amsterdam: Academic Press.

Jadidi, M., S. Moghtadernejad, and A. Dolatabadi. 2015. "A Comprehensive Review on Fluid Dynamics and Transport of Suspension/Liquid Droplets and Particles in High-Velocity Oxygen-Fuel (HVOF) Thermal Spray." *Coatings* 5, no. 4 (October): 576–645. https://doi.org/10.3390/coatings5040576.

Kantzas, A., J. Bryan, and S. Taheri. "Permeability: Harmonic-Average Permeability." *Fundamentals of Fluid Flow in Porous Media* (blog). November 2, 2014. https://perminc.com/resources/fundamentals-of-fluid-flow-in-porous-media/chapter-2-the-porous-medium/permeability/harmonic-average-permeability/.

Kitamura, K., M. Takahashi, K. Mizoguchi, K. Masuda, H. Ito, and S.-R. Song. 2010. "Effects of Pressure on Pore Characteristics and Permeability of Porous Rocks as Estimated from Seismic Wave Velocities in Cores from TCDP Hole-A." *Geophysical Journal International* 182, no. 3 (September): 1148–60. https://doi.org/10.1111/j.1365-246X.2010.04694.x.

Klatt, C. "Christian Klatt—Heating (in) Grinders." YouTube video, March 1, 2016. https://www.youtube.com/watch?v=XmtuuT36zFU.

Klatt, C. "Update on Grinding with Christian Klatt." YouTube video, November 17, 2017. https://www.youtube.com/watch?v=Pxaj4IEhaSs.

Knight, J. B., Jaeger H. M., and R. Nagel. 1993. "Vibration-Induced Size Separation in Granular Media: The Convection Connection." *Physical Review Letters* 70, no. 24 (June): 3728–31. https://doi.org/10.1103/PhysRevLett.70.3728.

Koekemoer, A., and A. Luckos. 2015. "Effect of Material Type and Particle Size Distribution on Pressure Drop in Packed Beds of Large Particles: Extending the Ergun Equation." *Fuel* 158 (October): 232–38. https://doi.org/10.1016/j.fuel.2015.05.036.

Labuza, T. P., C. Cardelli, B. Anderson, and E. Shimoni. 2001. *19th International Scientific Colloquium on Coffee: Physical Chemistry of Roasted and Ground Coffee: Shelf Life Improvement for Flexible Packaging, Trieste, Italy, May 14-18, 2001.* Paris: ASIC—Association for Science and Information on Coffee Proceedings.

Lindsay, S. 2009. *Introduction to Nanoscience.* Oxford: Oxford University Press.

Lingle, T. R. 1996. *The Coffee Brewing Handbook: A Systematic Guide to Coffee Preparation.* Santa Ana: Specialty Coffee Association of America.

Loeza, J. "Grinding Set." *The CoffeeSmith* (blog). May 8, 2015. https://sites.google.com/site/thegrindngo/components/grinding-set.

Malherbe, V. "Nesting Without Laying a Finger." *Café et Curiosité* (blog). January 24, 2020. http://ma.lher.be/2020/01/24/nesting-without-laying-a-finger/.

Marin, K. O., T. Požrl, E. Zlatić, and A. Plestenjak. 2008. "A New Aroma Index to Determine the Aroma Quality of Roasted and Ground Coffee During Storage." *Food Technology and Biotechnology* 4 (October): 442–7.

McKeogh, E. J., and D. A. Ervine. 1981. "Air Entrainment Rate and Diffusion Pattern of Plunging Liquid Jets." *Chemical Engineering Science* 36, no. 7: 1161–72. https://doi.org/10.1016/0009-2509(81)85064-6.

Michaud, D. "Surface Tension and Salts in Solution." *911 Metallurgist* (blog). February 11, 2016. https://www.911metallurgist.com/blog/surface-tension-salts-solution.

Moffat, R. *A Laboratory Study of Particle Migration in Cohesionless Soils.* Master's thesis, University of British Colombia, 2002.

Moroney, K. M., K. O'Connell, P. Meikle-Janney, S. B. G. O'Brien, G. M. Walker, and W. T. Lee. 2019. "Analysing Extraction Uniformity from Porous Coffee Beds Using Mathematical Modelling and Computational Fluid Dynamics Approaches." *PLoS One* 14, no. 7 (January): 1–24. https://doi.org/10.1371/journal.pone.0219906.

Motoyoshi, U. 2019. *The @wastingcoffee Guide to not Wasting Coffee.* Milan: Faema.

Nave, R. "Bulk Elastic Properties." *Hyperphysics* (blog). October 26, 2007. http://hyperphysics.phy-astr.gsu.edu/hbase/permot3.html.

Nielsen, L. C., I. C. Bourg, and G. Sposito. 2012. "Predicting CO_2–Water Interfacial Tension Under Pressure and Temperature Conditions of Geologic CO_2 Storage." *Geochimica et Cosmochimica Acta* 81, no. 1 (March): 28–38. https://doi.org/10.1016/j.gca.2011.12.018.

Oberg., E, D. J. Jones, L. H. Holbrook, H. H. Ryffel, and C. McCauley. 2016. *Machinery's Handbook.* South Norwalk: Industrial Press.

Ochi, J., and J.-F. Vernoux. 1998. "Permeability Decrease in Sandstone Reservoirs by

Fluid Injection: Hydrodynamic and Chemical Effects." *Journal of Hydrology* 208, no. 3–4 (July): 237–248. https://doi.org/10.1016/S0022-1694(98)00169-3.

Oliveros, N. O., J. A. Hernández, F. Z. Sierra-Espinosa, R. Guardián-Tapia, and R. Pliego-Solórzano. 2017. "Experimental Study of Dynamic Porosity and its Effects on Simulation of the Coffee Beans Roasting." *Journal of Food Engineering* 199 (April): 100–12. https://doi.org/10.1016/j.jfoodeng.2016.12.012.

Ozdemir, O., M. S. Çelikb, Z. S. Nickolovc, and J. D. Miller. 2007. "Water Structure and Its Influence on the Flotation of Carbonate and Bicarbonate Salts." *Journal of Colloid and Interface Science* 314, no. 2 (October): 545–51. https://doi.org/10.1016/j.jcis.2007.05.086.

Palmer, A. C., and T. J. O. Sanderson. 1991. "Fractal Crushing of Ice and Brittle Solids." *Proceedings: Mathematical and Physical Sciences* 433, no. 1889 (June): 469–77.

Park, S. K., and J. Y. Hu. 2010. "Assessment of the Extent of Bacterial Growth in Reverse Osmosis System for Improving Drinking Water Quality." *Journal of Environmental Science and Health, Part A* 45, no. 8 (May): 968–77. https://doi.org/10.1080/10934521003772386.

Perger, M. "Espresso Recipes: Strength." *Barista Hustle* (blog). January 30, 2017. https://www.baristahustle.com/blog/espresso-recipes-strength/.

Perger, M. "DIY Water Recipes: The World in Two Bottles." *Barista Hustle* (blog). May 1, 2017. https://www.baristahustle.com/blog/diy-water-recipes-the-world-in-two-bottles/.

Perger, M. "Sifting Coffee Grinds—An Experiment." YouTube video, January 3, 2019. https://www.youtube.com/watch?v=kI3zOwFG9mg.

Perger, M. "Coffee Grinder RPM." *Barista Hustle* (blog). January 4, 2020. https://www.baristahustle.com/blog/coffee-grinder-rpm/.

Petracco, M., and F. S. Liverani. 1993. *15th International Scientific Colloquium on Coffee: Espresso Coffee Brewing Dynamics: Development of Mathematical and Computational Models, Montpellier, France.* Paris: ASIC—Association for Science and Information on Coffee Proceedings.

Qu, X. L., L. Khezzar, D. Danciu, M. Labois, and D. Lahekal. 2011. "Characterization of Plunging Liquid Jets: A Combined Experimental and Numerical Investigation." *International Journal of Multiphase Flow* 37, no. 7 (September): 722–31. https://doi.org/10.1016/j.ijmultiphaseflow.2011.02.006.

Rao, S. 2008. *The Professional Barista's Handbook.* Self-published.

Rao, S. 2010. *Everything but Espresso.* Self-published.

Rao, S. 2013. *Espresso Extraction: Measurement and Mastery.* Self-published.

Rao, S. 2014. *The Coffee Roaster's Companion.* Self-published.

Rao, S. "Scott Rao—On Extraction Measurements." YouTube video, February 2, 2016. https://www.youtube.com/watch?v=tiqjf8-yuR4.

Rao, S. "Fines: Fine for Espresso, not so Fine for Filter." *Scott Rao* (blog). August 31, 2017. https://www.scottrao.com/blog/2017/8/27/fines-fine-for-espresso-not-so-fine-for-filter.

Rao, S. "Using Extraction Levels to Rate Grinders." *Scott Rao* (blog). October 30, 2018. https://www.scottrao.com/blog/2018/10/4/using-extraction-levels-to-rate-grinders.
Rao, S. *Instagram* (blog). February 4, 2020. https://www.instagram.com/p/B8K9yrbpUAh.
Rao, S. 2020. *Coffee Roasting: Best Practices.* Self-published.
Rhinehart, S. "Learn: What's the Deal with False Burrs?" *Prima-Coffee* (blog). December 16, 2016. https://prima-coffee.com/learn/article/grinder-basics/learn-whats-deal-false-burrs/32642.
Rober, M. "Liquid Sand Hot Tub—Fluidized Air Bed." YouTube video, November 28, 2017. https://www.youtube.com/watch?v=My4RA5I0FKs.
Rosti, M. E., S. Pramanik, L. Brandt, and D. Mitra. 2020. "The Breakdown of Darcy's Law in a Soft Porous Material." *Soft Matter* 16, no. 4 (January): 939–44. https://doi.org/10.1039/C9SM01678C.
Schenker, S., S. Handschin, B. Frey, R. Perren, and F. Escher. 2000. "Pore Structure of Coffee Beans Affected by Roasting Conditions." *Journal of Food Science* 65, no. 3 (April): 452–7. https://doi.org/10.1111/j.1365-2621.2000.tb16026.x.
Shonibare, O. Y., and K. E. Wardle. 2015. "Numerical Investigation of Vertical Plunging Jet Using a Hybrid Multifluid–VOF Multiphase CFD Solver." *International Journal of Chemical Engineering* 2015 (July): 925639. https://doi.org/10.1155/2015/925639.
Smith, B. T. 2015. *Remington Education: Physical Pharmacy.* London: Pharmaceutical Press.
Smrke, S. "A Science Guide to Coffee Freshness." *European Coffee Trip* (blog). March 19, 2020. https://europeancoffeetrip.com/coffee-freshness-science/.
Smrke, S., M. Wellinger, T. Suzuki, F. Balsiger, S. E. W. Opitz, and C. Yeretzian. 2018. "Time-Resolved Gravimetric Method to Assess Degassing of Roasted Coffee." *Journal of Agricultural and Food Chemistry* 66, no. 21 (May): 5293–300. https://doi.org/10.1021/acs.jafc.7b03310.
Socratic. *Instagram* (blog). December 3, 2015. https://www.instagram.com/p/-1SO4y-SuL8.
Sparks, T., and G. Chase. 2016. *Filters and Filtration Handbook.* Amsterdam: Elsevier.
Stanley, H. E., A. D. Araújo, U. Costa, and J. S. Jr. Andrande. 2003. "Fluid Flow Through Disordered Porous Media." *Fractals* 11, no. supp01 (February): 301–12. https://doi.org/10.1142/S0218348X03001963.
Sunergos Coffee. "Coffee Burr Grinder Overview." YouTube video. July 24, 2012. https://www.youtube.com/watch?v=-m3j7V5IlaA.
Uman, E., M. Colonna-Dashwood, L. Colonna-Dashwood, M. Perger, C. Klatt, S. Leighton, B. Miller, K. T. Butler, B. C. Melot, R. W. Speirs, and C. H. Hendon. 2016. "The Effect of Bean Origin and Temperature on Grinding Roasted Coffee." *Scientific Reports* 6 (April): 24483. https://doi.org/10.1038/srep24483.
Vaast, P., B. Bertrand, J.-J. Perriot, B. Guyot, and M. Genard. 2005. "Fruit Thinning and Shade Improve Bean Characteristics and Beverage Quality of Coffee (Coffea Arabica L.) Under Optimal Conditions." *Journal of the Science of Food and Agriculture* 86, no. 2 (October): 197–204. https://doi.org/10.1002/jsfa.2338.

van Heerwaarden, C. "Numerical Simulation of Rayleigh-Taylor Instability II." Vimeo (blog). January 21, 2014. https://vimeo.com/84675235.
Vargaftik, N. B., B. N. Volkov, and L. D. Voljak. 1983. "International Tables of the Surface Tension of Water." *Journal of Physical and Chemical Reference Data* 3: 817–20. https://doi.org/10.1063/1.555688.
von Blittersdorff, M., and C. Klatt. 2017. "Chapter 13—The Grind—Particles and Particularities," in *The Craft and Science of Coffee*, ed. B. Folmer, 311–28. Amsterdam: Academic Press.
Weber, D. "Blind Shaker." *Weber Workshops* (blog). September 11, 2019. https://weberworkshops.com/products/blind-shaker.
Wellinger, M., and C. Yeretzian. "Water: Why Quality Matters." *Café Europa | Autumn 2015* (blog). September 6, 2015. https://issuu.com/crimsonmedia/docs/caf___europa-autumn_2015/20.
Wellinger, M., S. Smrke, and C. Yeretzian. 2017. "Chapter 16—Water for Extraction—Composition, Recommendations, and Treatment," in *The Craft and Science of Coffee*, ed. B. Folmer, 381–98. Amsterdam: Academic Press.
Wellinger, M., S. Smrke, and C. Yeretzian. 2018. *The SCA Water Quality Handbook*. Santa Ana: Specialty Coffee Association.
Worku, M., B. de Meulenaer, L. Duchateau, and P. Boeckx. 2018. "Effect of Altitude on Biochemical Composition and Quality of Green Arabica Coffee Beans Can Be Affected by Shade and Postharvest Processing Method." *Food Research International* 105 (March): 278–85. https://doi.org/10.1016/j.foodres.2017.11.016.
Yam, K. L., ed. 2009. *The Wiley Encyclopedia of Packaging Technology*. New Jersey: Wiley.
Yang, L., D. Gao, Y. Zhang, J. Tang, and Y. Li. 2019. "Relationship Between Sorptivity and Capillary Coefficient for Water Absorption of Cement-Based Materials: Theory Analysis and Experiment." *Royal Society Open Science* 6, no. 6 (June): 190112. https://doi.org/10.1098/rsos.190112.
Yang, Y., F. D. Siqueira, A. S. L. Vaz, Y. Zhenjiang, and P. Bedrikovetsky. 2016. "Slow Migration of Detached Fine Particles Over Rock Surface in Porous Media." *Journal of Natural Gas Science and Engineering* 34 (August): 1159–73. https://doi.org/10.1016/j.jngse.2016.07.056.
Yang, Z., J. An, A. Wang, and Y. Fan. 2013. *Sixth International Conference on Advanced Computational Intelligence (ICACI): A Novel Security Method in Wireless Network Based on Percolation, Hangzhou, China, October 19–21, 2013*. Piscataway: Institute of Electrical and Electronics Engineers. https://doi.org/10.1109/ICACI.2013.6748500.
Yeh, T.-C. J., M. Ye, and R. Khaleel. 2005. "Estimation of Effective Unsaturated Hydraulic Conductivity Tensor Using Spatial Moments of Observed Moisture Plume." *Water Resources Research* 41, no. 3 (March): W03014. https://doi.org/10.1029/2004WR003736.

Yin, Z., Q. Jia, Y. Li, Y. Wang, and D. Yang. 2018. "Computational Study of a Vertical Plunging Jet into Still Water." *Water* 10, no. 8 (July): 989. https://doi.org/10.3390/w10080989.

Zhong, W., F. Yue, and A. Ciancio. 2018. "Fractal Behavior of Particle Size Distribution in the Rare Earth Tailings Crushing Process Under High Stress Condition." *Applied Sciences* 8, no. 7 (June): 1058. https://doi.org/10.3390/app8071058.

Zhu, S.-Y., Z.-M. Du, C.-L. Li, A. Salmachi, X.-L. Peng, C.-W. Wang, P. Yue, and P. Deng. 2018. "A Semi-Analytical Model for Pressure-Dependent Permeability of Tight Sandstone Reservoirs." *Transport in Porous Media* 122, no. 2 (January). https://doi.org/10.1007/s11242-018-1001-x.

Zięba, Z. 2017. "Influence of Soil Particle Shape on Saturated Hydraulic Conductivity." *Journal of Hydrology and Hydromechanics* 65, no. 1 (March): 80–87. https://doi.org/10.1515/johh-2016-0054.

Index

Jonathan Gagné is a researcher in astrophysics at Montréal's Planétarium Rio Tinto Alcan and is an adjunct professor at Université de Montréal, specializing in the study of brown dwarfs, exoplanets and young stars near the Sun. His passion for coffee led him to write a blog about the technical aspects of coffee brewing at http://coffeeadastra.com.